SALTS MILL

SALTS MILL
THE OWNERS AND MANAGERS 1853 TO 1986

MAGGIE SMITH & COLIN COATES

First published 2016

Amberley Publishing
The Hill, Stroud,
Gloucestershire, GL5 4EP

www.amberley-books.com

Copyright © Maggie Smith and Colin Coates, 2016

The right of Maggie Smith and Colin Coates to be identified as the Authors of this work has been asserted in accordance with the Copyrights, Designs and Patents Act 1988.

All rights reserved. No part of this book may be reprinted or reproduced or utilised in any form or by any electronic, mechanical or other means, now known or hereafter invented, including photocopying and recording, or in any information storage or retrieval system, without the permission in writing from the Publishers.

ISBN: 978 1 4456 5753 0 (print)
ISBN: 978 1 4456 5754 7 (ebook)

British Library Cataloguing in Publication Data.
A catalogue record for this book is available from the British Library.

Typeset in 11pt on 14pt Celeste.
Typesetting by Amberley Publishing.
Printed in the UK.

Contents

Author's Note and Acknowledgements	7
Foreword by Nick Salt	8
Foreword by Jamie Roberts	11
1 Sir Titus Salt 1853–1876	13
2 The Salt Descendants and the Stead Family 1876–1893	30
3 Sir James Roberts and the Consortium Years 1893–1918	55
4 The Syndicate Years 1918–1957	80
5 The Illingworth Morris Years 1957–1986, and the End of Textile Production at Salts Mill	103
Postscript: Salts Mill 1987 to the Present	123
Bibliography	125

Author's Note and Acknowledgements

This short account of the successive owners and managers of Salts Mill could not have been produced without thorough and intensive new research undertaken by Colin Coates, a fellow historian from Saltaire. Others who deserve many thanks for the provision of their own work, their files and many helpful observations and corrections are Dave Shaw, David King, Ian Watson, and especially Sandi Moore for her help in finding some key images and texts from within the Saltaire Archive. I owe particular thanks to Nick Salt and Jamie Roberts, descendants of the 'two Lords of Saltaire' for reading chapters two and three (which are about their ancestors) so carefully, and for providing corrections to these. Also to Anna Roberts, Strathallan Castle, for her loan of many important documents and images about the Roberts family.

I owe a large debt of gratitude to Mark Keighley for his kind permission to use aspects of his important book about Bradford's textile industries, *Wool City*, published in 2007. I am very glad to have met Donald Hanson in 2012 and to have worked with him for over twelve months to produce his personal account of his time as an employee who became the Chair and joint Chief Executive of Illingworth, Morris, in 1981 – Chapter 5 could not have been written without his major contribution.

I accept full responsibility for any errors in the text and must also apologise to the families of many of the important people who participated in the steering of the companies after 1918, but for whom this book did not have the space with which to detail their important contributions. My greatest wish, however, is for this small book to encourage new in-depth research into some of the people who are sketchily outlined in this book, and into the many others whom I did not have the space or time to research and write about.

Maggie Smith
2016

Foreword by Nick Salt

Titus Salt Jr

Titus Salt Jr was my great-grandfather, and always a bit of an enigma to me and my siblings. Our father John Salt had died young, meaning that our main connection with the Salt family and its history was through his mother (not a Salt) and his two sisters. They lived in The Old Rectory at Thorp Arch, the house bought by Catherine Salt, the widow of Titus Jr, a few years after she had left Milner Field in 1903. Her husband had died in 1887, and she had lived on with her four children in the grand house that they had built together in 1872, before selling it to James Roberts, the new owner of the mill and village at Saltaire. The Old Rectory and much of its contents were passed on to her son Gordon Salt, our grandfather – and so our father was brought up there.

As I got older I became aware that the house was largely furnished with the possessions that Catherine had brought with her from Milner Field, which helped to provide us with a direct link to the very different lives of Titus Jr and his family, way back in the heady days of Saltaire a century earlier. It was unfortunate that many of those possessions got scattered when The Old Rectory was eventually sold in 1978, but we still have, or have access to, some of them, including furniture, china, glassware, jewellery, photographs and documents. I still enjoy using his mahogany workbench and rosewood-handled woodworking tools, stamped with his name. Many of these items are now safely lodged in such places as the Saltaire Archive at Shipley College (including documents and photographs), the Bradford Industrial Museum (mostly photographs), and Lotherton Hall (furniture and textiles), where they continue to provide valuable original sources of information for a keen group of local historians.

The Salt family, ever since the era of Saltaire, has not been very prolific. Sir Titus and Caroline Salt had a large family of eleven children, seven of whom either died young, did not marry, or who married and had no children. At the age of fifty,

Titus, instead of relaxing into a comfortable retirement, had used his experience, skills and wealth to establish the pioneering Saltaire mill and village. His two principal aims at that time were to provide better working and living conditions for his many workers and their families, and 'to leave something for his boys to do' in providing employment for his five surviving sons. The first of these aims was clearly achieved, but the second was less successful, and the Salt dynasty at Saltaire only lasted about forty years. Two of his sons had very little involvement with the mill and the village and, of the other three, it was mainly Titus Jr, the youngest, who seemed to have the enthusiasm and vision to carry on his father's work, both at Saltaire and out in the wider community.

Over the years I have become increasingly interested in the busy life of Titus Jr, but it was only when reading the draft of this book that I became fully aware of the wider extent of his many activities. It is impressive how much he had managed to achieve by the time of his early death of a heart attack at the age of forty-four, and I have a lot of respect for how well and wisely he had used the advantages

Princess Beatrice and Prince Henry of Battenburg in the grounds of Milner Field. Titus Junior looks ill.

and privileges of his family's wealth and status. But his close involvement in the management of the Salt family empire, with its inevitable ups and downs, must have been a great strain upon his mental and physical health. In the group photograph of the royal visit to Saltaire in May 1887, he looks ill and stressed, and he died only six months later.

It was said at the time that, 'He had no mercy on himself ... He was never really happy unless he was at work and he always worked hard ... He did, however, delight in his gardens, and his extensive glass houses at Milner Field were admirably designed and managed, all by himself.'

I am grateful to Maggie Smith and Colin Coates for unearthing so much fascinating new information about him, and hope that their book will inspire others to do the same.

Nick Salt

Foreword by Jamie Roberts

In my childhood home there was a large portrait as you went up the stairs: an unsmiling man with a firm gaze, whose thick red robes and white handlebar moustache made him resemble a high court judge. As I found out many years later, the man in the picture – who was in fact receiving an honorary doctorate – was my great-great-grandfather, Sir James Roberts.

James Roberts became part of the story of Salts Mill when it was facing its first and arguably greatest existential crisis. The previous owner had been declared bankrupt and several thousand workers – in effect the whole town of Saltaire – faced a bleak future. It is a testament to Roberts's vision and (that much-coveted Yorkshire quality) bloody-mindedness that, by the end of his tenure in 1918, the mill had not just been saved but was again one of the great industrial powerhouses in the world.

For a long time, the standard-issue histories of Saltaire contained very little about James Roberts. It was as if the passing of the great Salt era at the mill in 1892 had given way to a long interregnum of non-descript business partnerships until the renaissance under Jonathan Silver in the late 1980s. The situation wasn't helped by the fact that Roberts himself appears to have chosen to avoid the limelight – he never changed the name of the mill, erected statues or named streets after himself. It was only too easy to overlook his immense contribution to the story of Saltaire. Thankfully this has started to change in the past decade or so, largely thanks to the dedicated research of the members of the Saltaire History Club.

My own interest in Saltaire began with the arrival of the Hockney gallery in Salts Mill – which felt and still feels like a breath of fresh air among the industrial heartlands of West Yorkshire. This personal connection to Salts Mill soon branched out into a need to find out more about the ancestor in the portrait who called this great mill and town his own for a quarter of a century. In recent years I have been lucky enough to be part of the project to understand more about his life, and to meet representatives of the Salt and Silver families with whom I have so much shared history.

I am pleased to see books such as this one adding new layers to what we understand about the great enterprise and settlement of Saltaire and its mill.

Jamie Roberts

1

Sir Titus Salt 1853–1876

Sir Titus Salt (1803–1876): Founder of Salts Mill and Saltaire

To gain an insight into the man who built Salts Mill and founded Saltaire, it is important to understand how he came to be involved in textile manufacture and what influences would have shaped his personal and business life.

The most authoritative work on Titus Salt, *The Great Paternalist; Titus Salt and the Growth of Nineteenth-Century Bradford* (J. Reynolds, 1983) provides a thoroughly researched work exploring the social, political and economic changes of Salt's time. Reynolds takes the starting point for his work as 1847, noting that the populated areas of what was to eventually become the City of Bradford involved four townships: Bowling, Heaton, Manningham and Bradford itself. Its economy then was based on a mixture of farming and manufacture, due to an important change in the late seventeenth century when worsted stuff manufacture was introduced to the west of the West Riding from East Anglia.

Worsted cloth was different to woollen cloth, using different types of yarn. The long wool fibres were spun together along the axis of the yarn, creating a smoother cloth that enabled pattern definition. The separation of long fibres (tops) was achieved by combing the wool with heated, heavy metallic combs, leaving short fibres (noils) that were sold to wool manufacturers. Wool combing and weaving were almost entirely domestic industries that were well established by the mid-nineteenth century in the area of the townships and, while there were several factories in the parish of Bradford, there were none in the town itself at this time.

The first significant change in the parish of Bradford had occurred in the eighteenth century with the establishment of iron manufacture in Bowling and Low Moor; there were plentiful coal deposits in these areas, providing tools and fuel for the emerging textile industry. By 1780, the local population involved in worsted manufacture had risen from 25 per cent to between 45 per cent and 50 per cent. At the turn of the century, the bulk of the spinning processes were carried out

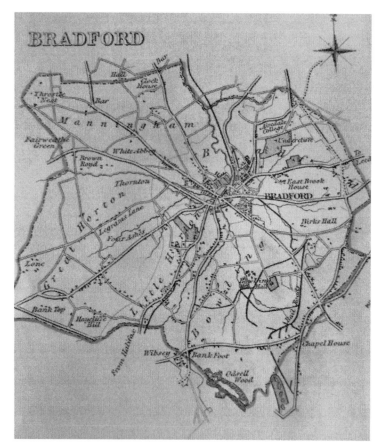

Left: Creighton's 1835 Map of Bradford. Note the evidence of large rural areas.

Below: Artist's depiction of Bradford in 1873.

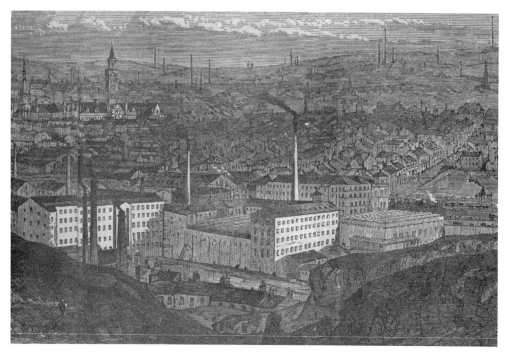

in factories that employed women and children to do the work. The power loom was introduced in 1824 and experiments with machine wool combs had begun. The wool market itself moved from Wakefield to Bradford and leadership of the Bradford textile industry was firmly in the hands of manufacturers and merchants. Reynolds (1983) notes,

> There were few restraints on entrepreneurs, few social demands on profit, low rates for the parish, public space available at low cost, no regulations on hours of work and the capital cost of expansion was very low.

Thus by the 1830s Bradford had become a magnet for people seeking work or business opportunities, experiencing a further population growth of 78 per cent in the first third of the nineteenth century. There was a well-established upper class of conservative, Anglican gentlemen, but a new middle class had begun to emerge among well-to-do farmers, important manufacturers, and merchants. The economic interests of both groups were similar, but religious affiliations differed, as did political affiliations. The new middle class were often 'dissenters' in terms of religion and liberal in their political views.

This exponential population growth and continuously increasing mechanisation of worsted manufacture had a number of social and moral impacts that were of concern to both the upper and middle classes. The townships moved rapidly from relatively small communities where people living in the areas had strong familial roots and communal networks. In itself, this meant that individuals and families were known by and knew many others living and/or working nearby. This provided some degree of 'social control', in that any wrong-doing would be noticed and subject to communal censure. The advent of a population of strangers – both to the indigenous population and to each other – appeared to create a 'licence' for immoral and unwelcome behaviours that were much more difficult to counteract with community pressures.

In addition, some of the 'masters' in the textile crafts, who had 'put work out' to home-based workers, gradually lost their prior status and resented this. In fact, the changes to the area created a new 'working class' who had very hard times when work was short. This in itself brought active involvement for some working people, in the formation of trade unions and allegiances to the radical Chartist movement. Chartism was a national working-class movement, which emerged in 1836 and was most active between 1838 and 1848. The aim of the Chartists was to gain political rights and influence for the working classes. Chartism got its name from the formal petition, or 'People's Charter', that listed the six main aims of the movement:

- All men to have the vote
- Voting to take place by secret ballot
- Parliamentary elections to be held every year, not every five years

- Constituencies to be of equal size
- Members of Parliament to be paid
- The property qualification for becoming a Member of Parliament to be abolished

The origins of Chartism were complex and, for William Lovett, peaceful persuasion by respectable working men – 'moral force' – was the best way to win the points in 'the Charter'. This strategy clashed with that of Feargus O'Connor, a charismatic speaker who used mass meetings and the widely read *Northern Star* to unite the forces of the working class behind him. His popularity was immense and he advocated the use of physical force to fight for the charter. Chartism was a national movement, but it was particularly strong in the textile towns of Lancashire and Yorkshire, as well as other areas where rapid industrialisation created large working-class populations.

The movement was formed not very long after the French Revolution, which had similarities in uniting the poor and the middle classes. Its existence was certainly of concern to the property-owning men of both conservative Anglican or Liberal/Radical background. The difference between the two classes, however, was important and many among the emerging middle classes had sympathy with 'moral force Chartism'.

It was at the beginning of this transformational period that the Salt family arrived in Bradford. Daniel Salt brought his family to the town probably in 1820

A Chartist gathering, Kennington Common, London, 1848.

or 1821, when he was aged thirty-eight or thirty-nine years. He had considerable experience of business and had been an iron founder, a dry salter and a white cloth merchant; he also begun farming in 1811. He had been attracted by Bradford's expanding economy and sold his farm to invest the capital in the wool industry. He took an office in the centre of Bradford and established a wool-stapling business – buying wool from suppliers, sorting and grading this and then selling it on for spinning and weaving. He also bought land in Cheapside and built first one and then a second warehouse in Bradford that became his headquarters.

He was a West Riding man with useful connections, and was well established in the dissenting community; a Congregationalist in faith. He rapidly immersed himself in public life in Bradford. In 1824 he was appointed to the vestry office as an overseer of the poor; in 1825 he was named as a commissioner; in 1831 he was elected as a constable of the manor and ardently supported political reform for the emancipation of Catholics, the abolition of slavery and demand for 'Free Trade'. In 1837 he was elected to the first Board of Guardians and from 1835 to 1841 was prominent in radical organisations. He was a tall, well-made man with a reputation for intelligence and wit and, despite his faith, was not known to be a rigid puritan. He had a well-pronounced stammer that did not seem to affect his progress in either business or public life.

He had eight children with his wife Grace Smithies; Hannah Marie (b. 1806) and Hannah (b. 1821) died in infancy, while Isaac Smithies (b. 1810) died at the age of nine years. Five children survived, and Daniel had built the family a home in Manningham by 1834. His wife Grace was known to be devoutly religious, and she gave all the Salt children their religious education. The oldest son Titus was nineteen years of age when the family arrived in Bradford; sisters Sarah, Ann and Grace were eighteen, four and ten years of age and brother Edward was eight years old.

Titus, born in 1803, was initially educated in a 'dame school' in Morley (near Leeds), and then at Batley Grammar School; he was later given a plain commercial education at a Wakefield school run by a Mr Enoch Harrison. He was described as tall and stout, with a heavy appearance, and as having a studious turn of mind, rarely mixing with his school fellows and being somewhat introverted. He was known to have some difficulty in reaching decisions and was unable to speak confidently or coherently in public. He had a highly developed sense of responsibility, however, and this offset his diffidence.

In 1820 Titus had been apprenticed to a wool stapler, a Mr Jackson of Wakefield, and when the family came to Bradford he spent two years with William Rouse & Son, where he gained a more comprehensive insight into the textile business. In 1824 he joined his father, and the firm became Daniel Salt & Son. Titus worked in the family firm for ten years, during which time the firm began to specialise in Donskoi wools from Eastern Europe. He took part in public life alongside his father and was also a firmly committed liberal-radical dissenter. In 1826 he was enrolled as a special constable during rioting at Horsfall's mill when a very bitter strike took place in the local worsted industry. This perhaps indicated both his own strong concerns about worker unrest and his concern to sustain the rights of employers. At this time, he was also appointed as superintendent at Horton Lane Chapel Sunday School, although he was too timid to read the prayers to the congregation.

In 1830 he married Caroline Whitlam of Grimsby, the youngest daughter of Mr George Whitlam, who was a wealthy sheep farmer. He first took a house in North Parade, Bradford, very near to his parents' house but, between 1836 and 1843, he lived at the junction of Thornton Road and Little Horton Lane. This was very near Nelson Court and Fawcett Court, the heart of the worst slum property in Bradford, and his understanding of the deprivation of many families will have developed here. His eldest children spent their early years living very close to this part of town. William Henry was born in 1831; George in 1833; Amelia, 1835; Edward, 1837; Herbert, 1840; Fanny, 1841; and Titus, 1843.

Salt and his father had started a spinning department of their wool-stapling business, using rooms in Thompson's Mill at Goitside Thornton Road, Bradford. They had been joined in the business by Edward Salt in 1834 and, shortly after this, Daniel Salt retired. Within a few months Titus had also left the firm and started to work on his own account (the family partnership was dissolved in 1835).

Sir Titus Salt.
(© The Saltaire Archive)

Lady Caroline Salt in later years. (© The Saltaire Archive)

Titus Salt started his new venture at Hollings Mill and quickly acquired premises on or near Hope Street. In 1836 he took over a large mill in Union Street (whose previous owner had been Daniel Illingworth) and this became the headquarters of his rapidly expanding alpaca manufacturing. He also employed large numbers of hand woolcombers in the Manchester Road area and hand weavers in the nearby villages of Allerton, Clayton and Baildon. His success was aided by newly acquired capital from his father's will, and he was to add three more Bradford mills to his business before the mid-1840s.

The wealth from his business supported his very active public career and, at one time or another, he occupied almost every local public office. He was elected High Constable of the Manor in 1841/42, organising and presiding at all town meetings requisitioned by the required number of rate payers. From 1847 to 1854, he was the senior alderman in the new Bradford Corporation for South Ward. In 1848/49 he was elected mayor of the town and presided at almost all full council meetings. He also sat regularly on the bench of the Bradford Court and was an active member of the Chamber of Commerce. He was always a member of the inner circle of well-to-do radical dissenters who played a leading role in the reformation of Bradford's public life. Meanwhile, through the forties and fifties he was particularly active in politics, supporting the cause of the Anti-Corn Law League and showing much sympathy with the ideals of 'moral force' Chartism. While demonstrating a view that working people should be loyal to their employers, he clearly sympathised with the plight of the growing urban working class in that they had somewhat precarious incomes with which to care for themselves and their families, while having little access to decent homes.

Although it had become a dominant area in the wool trade, Bradford had been slow to develop any innovations in textile manufacture during the period in which the Salt family lived there. Success had come about through the adaptation of devices and inventions in machinery. After 1836 the production of mixed rather than pure woollen fabrics had commenced, wherein cotton and silk were used with wool and mohair – innovation driven by difficulties in obtaining sufficient quantities of long-fibre wools. At the same time the manufacturers created a demand for lighter and more elegant fabrics. Use of a wider range of materials also cheapened the costs of production. By 1837 Bradford mills were using cotton as warps, but what was to become more important was the development of cloth that utilised the long, shiny hair of the alpaca.

Alpaca, imported from Peru, became known in the United Kingdom in the early years of the nineteenth century, when new trading connections were being developed with Latin America. Although several amateur innovators experimented with the new material, it was not until the 1830s that its commercial properties were tested. Early trials in worsted manufacture with this material were

Alpacas.

problematic. The first attempts to produce cloth with it, carried out by Benjamin Outram of Halifax, involved very high costs and the finished cloth lacked lustre. A number of other firms also produced experimental materials but were not able to capitalise on its potential by discovering how best to comb and spin this new material.

Titus Salt, however, was no stranger to experimentation and had followed his father's lead in the use of Donskoi wool. He had the persistence and ingenuity to resolve the problems with alpaca wool. Salt was experimenting with alpaca wool by 1837 and there is documentary evidence in his small day book, kept between 1834 and 1837, wherein he recorded costs of the material and its processing. It was to be trials with alpaca weft and cotton warp that proved successful, and there are records of his early sales of the new cloth in 1839. Titus Salt was the only spinner of alpaca weft in Bradford. He had worked in great secrecy for eighteen months with trusted assistants, and the cloth he produced in this way was durable, relatively light, lustrous and reasonably priced. Salt and others also made greater and greater use of mohair from the Levant and small amounts of Australian wool.

By 1843 he had had enormous success and had ceased to live in Bradford, moving with his family to a handsome mansion called Crow Nest at Lightcliffe (near Halifax), 10 miles away from his business and his public commitments. By this point in time there were a number of Bradford textile manufacturers who, by any standards, were wealthy men. Some joined the ranks of the aristocracy and titled gentry, and a number confirmed their success in industry by converting their resources into landed estates. Samuel Cunliffe Lister bought the Masham and Jervaulx estates in North Yorkshire and, some years later, became Lord Masham. Among the non-titled, Mathew Thompson bought an estate in Kent; John Wood had many acres in Hampshire; and George Hodgson, the textile engineer, bought an estate in Norfolk.

In 1851, Henry Forbes estimated the value of the British textile trade at £12,500,000, of which the West Riding of Yorkshire accounted for £8,000,000. Despite this and the undoubted wealth of many of the Bradford manufacturers, few of the fortunes reached the highest level of wealth. Only one manufacturer of the time was listed as a millionaire at his death – William Foster of Black Dyke Mills.

Titus Salt, one of the wealthy of Bradford, did not choose to convert his fortune into a landed estate (that was to be the choice of his oldest son, William Henry, probably in 1865). Instead he had the originality of mind to go on to build his memorial to the future through the creation of an industrial community settlement, to be named Saltaire. The question as to why he should choose to do this and act differently to his peers has never been satisfactorily answered. The site he chose for his new mill was 4 miles away from Bradford in a rural area on the banks of the River Aire at Shipley.

Reynolds (1983) poses some probable practical reasons for siting his new mill there:

- The recent developments in woolcombing machinery allowed him to think in terms of total factory production rather than different mills and machinery performing different functions.
- To buy any quantity of new machinery required him to have a great deal more space than he then had across his six mills.
- To concentrate all his activities in one large mill would cut out waste and enhance supervision of related functions.
- It was impossible to find a suitably large site in Bradford and the site he selected was, on commercial grounds alone, one of the best that could be found in the north of England.
- The site stood between the River Aire and the Midland Railway line, and through its centre ran the Leeds/Liverpool canal. It was a site at the crossroads of the principal lines of communication, north/south and east/west, ensuring cheap transport for coal, raw materials and finished goods.

Known to be relatively indecisive, Salt nevertheless set out to build a grand 'vertical mill' – one that could take the raw materials in to the factory, as yet untreated, and have the machinery and capacity to perform the many processes required to complete and finish cloth of high quality. He commissioned the architects Lockwood & Mawson to design the mill, and settled on a fifteenth-century renaissance style. William Fairburn, a great engineer of the day, designed the layout of the interior and constructed the engineering facilities. The building of the mill commenced in 1851 and was completed in 1853. It had a production capacity

Artist's impression of Salts Mill, *c.* 1860. (© The Saltaire Archive)

for 30,000 yards of cloth to be completed every working day and employment at full capacity for 3,000 people.

Salt was aided in equipping his new mill by being one of only two men to get the better of Samuel Cunliffe Lister. Lister, another of Bradford's major cloth manufacturers, was a skilled inventor himself. He had developed a practice for buying the patents of all new textile machinery, particularly those involving the difficult mechanisation process for woolcombing. Salt collaborated with Edward Ackroyd, the Halifax worsted manufacturer, to buy the Heilman patents, which they both knew Lister needed to gain a monopoly in the manufacture of machine combs. They paid £33,000 for this but defeated Lister, much to his chagrin. As a result, Salt was able to stock the woolcombing shed at Salts Mill at a cost of £200 per machine rather than the £1,200 that was Lister's market price at the time. Reynolds (1983) recounts this as clear evidence of Titus Salt's business acumen and dogged determination to succeed.

The new mill had a grand opening on 20 September 1853, Salt's fiftieth birthday and an occasion of lavish hospitality. The business was to remain a family firm throughout Salt's lifetime. Salt brought in as junior partners two very able men from his senior staff in 1854: Charles Stead and William Evans Glyde (the brother of Jonathan Glyde, the Horton Lane Congregational pastor). His sons William Henry and George Salt entered the firm at the same time, and Edward Salt joined the firm around the time when Glyde decided to retire, probably in 1865. Titus remained as senior partner all his life, his position very clearly defined by the articles of association for the company.

His vision for his business was much greater than that of building a mill with an immense capacity for textile manufacture. From the outset, the plan was to also build an industrial settlement. Indeed, the first plan for the settlement to be named Saltaire was more sizeable in scope, as recorded in 1854 it:

> provided for a population of 9,000 to 10,000 and sites for a church, baths and washhouses, a Mechanics Institute, hotels, a covered market, schools and alms-houses, an abattoir and dining hall and a music room.

Salt began the arrangements for building the village as soon as the mill was completed and open. In October 1853, tenders were invited for the first fifty-three 'cottages' and, by October 1854, fourteen shops were ready for occupation, 163 houses and boarding houses had been completed, and around 1,000 people were in residence. The village was to develop around the Bradford–Keighley road, and house building was started in a series of rigidly defined parallelograms. Three types of accommodation were made available to fit the needs of particular families, either for the space they required or in relation to the family income or status within the village and the factory. Between 1866 and 1869 the first part of the housing programme was completed. The erection of important public

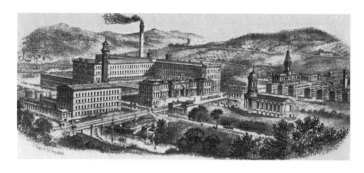

Artist's impression of Saltaire, c. 1880.

buildings had been accomplished on both sides of what had been named Victoria Road. A number of shops were included in the new streets, and at all levels the housing compared favourably with housing in Bradford. They were supplied with gas directly to each home from the mill, cheaper than that available locally, and initially water was also supplied directly by the mill. Each house had its own lavatory in the yard, and the streets had been laid out and drained before the houses were built.

Although most of the public buildings were erected during the final stages of development, some general amenities were available from the beginning. A dining room was built opposite the mill to provide cheap meals for workers travelling in to work. Space there was used for a regular programme of speakers and a factory school started life in the dining room also. The Saltaire Literary Society and Institute began in November 1855; a reading room was open every evening; and a debating society was established. In 1863 wash houses and baths were erected, although these did not prove popular with 'respectable women', who preferred to take a bath in their own homes. A smaller factory establishment was added in 1868 on the site of an old water mill at the river's edge (known then and now as 'new mill'). The factory school moved to new purpose-built premises further up Victoria Road. In 1868, forty-five almshouses were built at the southern end of Victoria Road and an infirmary and dispensary was opened on the opposite side of the road. In 1869, the year that Titus Salt was created a baronet, the building of an institute to provide further education and a social club commenced. A park was established across the river with land bought by Salt in 1870 and landscaped attractively by a Mr Gay of Bradford. The final houses in Dove Street and Jane Street were completed in 1875 and, by 1871, Saltaire already had a population of 4,390.

After 1861, however, Titus Salt lived in semi-retirement, relinquishing much of the day-to-day control of his grand mill and the settlement of Saltaire to his partners. His wife, Grace, had borne eleven children and, after 1843, the family lived firstly at Crow Nest, Lightcliffe, and between 1858 and 1868 at Methley Hall near Leeds. The family moved back to Crow Nest, which the family had left unwillingly in 1858, when Salt was able to buy it in 1868 and returned thankfully to what they had come to regard as their family home. Both Crow Nest and

Methley Hall were said to be 'furnished with all the elegance and luxurious taste wealth can command'. Balgarnie describes Crow Nest in some detail as a

> mansion of hewn stone consisting of a centre portion with a large wing on each side, connected by a suite of smaller buildings, in the form of a curve ... the south side presents a landscape of secluded beauty, in which wood and lake, lawns and terraces, flower gardens and statuary delight the eye.

From the time the family moved to Crow Nest, they were able to live the life of the wealthy upper class in luxurious surroundings and in contact with all the important people of the day. Titus Salt himself became a 'crested and carriage gentleman', but he didn't succumb entirely to the status he had achieved. Eight of his eleven children had survived. Whitlam and Mary had died in infancy from scarlet fever, and Fanny of tuberculosis at the age of twenty. Just two children were born at Crow Nest – the youngest daughters, Helen and Ada. William Henry and George first went to school at Huddersfield College, established in 1838 for the children of dissenter families who wanted their boys to have some business education and to avoid the lure of the public school. Here there was a particular emphasis on modern languages, writing, commercial education and book-keeping. Nevertheless, when Edward was ten years old, in 1847 he and his two older brothers went to Mill Hill school in London, which was a public school for the children of wealthy dissenters. The remaining two sons, Herbert and Titus, went to

Mill Hill School today.

Mill Hill when they were over ten years of age, but all the brothers left the school in 1855 due to a serious outbreak of scarlet fever there, and they did not return. Titus (Titus Junior as he was to become known) had only two years schooling at Mill Hill.

How the difference in the early lives of the male children affected their adult decisions with regard to carrying on the family business founded by their father is a matter of speculation. None of the sons had any formal education after the age of sixteen years (the girls were educated privately). All the boys worked in the mill, learning the business at first hand, and all, except Herbert, were accepted into partnership. They all experienced very different social and educational contexts to their father. The question as to how much his character and religious and political beliefs influenced them is a difficult one, given that Sir Titus Salt himself leaves few clues to his own personality and style of parenting.

Reynolds (1983) comments frequently about how Salt's reluctance to speak in public or to produce written materials left little for the historian to study in order to assess his business methods or character fully. There are few details of how he conducted his business in Bradford, but it is fairly clear that the preliminary processes in textile manufacture – wool sorting, and machine and hand washing – were controlled from Union Street, where administrative and financial matters were also dealt with. He is thought to have enjoyed the benefits that the mixed systems of production gave but, like others, he had to face the social and moral problems it presented in the new, highly urbanised society within Bradford. His attitude to factory reform, wage remuneration and the introduction of machinery was the standard one for men of his class and time. Wages found their own level and mechanisation could not be artificially controlled. He was not in favour of factory legislation limiting hours of work and, throughout the 1840s, he was chairman of the manufacturers' opposition committee.

Occasionally it was clear that he was prepared to provide entertainment and parties for his employees. In 1847 he celebrated the Liberal electoral victory with a couple of factory parties, one of which was a tea party and dance at the Odd Fellows Hall for 350 women weavers and their escorts. A newspaper report from six years earlier, however, had him bringing to trial two woolcombers suspected of stealing charcoal from him, which on investigation was not found to be the case. There were other incidents that show Salt as an employer applying the discipline of the trade as all did at the time, and there is no reason to suppose that Salt was anything other than a strong disciplinarian in matters concerning the work of his company. Nevertheless, Salt emerged from the difficult years of transition and immense growth in the trade with his reputation as an employer unsullied.

After building the magnificent mill at Saltaire and founding a village for his employees with better housing conditions than could be found anywhere for workers in Bradford, he began to withdraw from public life and duties. By 1853 he was a regular absentee from most of the corporation meetings he could have attended and

The raw materials and finished products were often transported by canal barge. (© Saltaire Archive)

let it be known in that year that he did not wish to be re-elected as an alderman. He had sat regularly on the bench of the Bradford Court, applying firm justice to recalcitrant textile workers but also offering personal help to men and women who had suffered harrowing misfortune; however, he ceased his role here in 1854.

He remained particularly active in politics in the 1850s and was ready to stand for Parliament in 1857; he took a seat in 1859 and remained in Parliament for two years. He found parliamentary life something of a burden and in 1861 resigned his seat. He was beginning to be troubled by gout, which was to plague him for the rest of his life. He was much more rarely seen in public and for his last years was to be much more involved with his charitable, philanthropic activities and the completion of Saltaire. He had become 'the grand old man' of Bradford liberalism and still made some political interventions in these later years, including a strong attack on W. E. Forster's education policies in the election of 1874. It is clear that his business, his family and his foundation of the industrial 'town' of Saltaire were always more important to him than politics or the public life. The questions remain, however, as to why he founded this lasting memorial, which is now a designated World Heritage Site. A number of academics, writers and others have tried to identify his motives for doing so at the age of fifty years, when the project began; this was a time when he had more than enough wealth to retire peacefully from business.

In David James's 2004 assessment, he comments,

> Salt's motives in building Saltaire remain obscure. They seem to have been a mixture of sound economics, Christian duty, and a desire to have effective control over his workforce. There were economic reasons for moving out of Bradford, and the village did provide him with an amenable, handpicked workforce. Yet Salt was deeply religious and sincerely believed that, by creating an environment where people could lead healthy, virtuous, godly lives, he was doing God's work.

Perhaps Reynolds (1983) gets nearer to the likely answer. He notes that Saltaire was one of several industrial settlements developed by great employers of labour in the mid-nineteenth century, and he disagrees with those historians who describe it as an example of paternalistic philanthropy:

> It was a company village subject to the control of an employer/landlord and it is primarily as an experiment in industrial relations that it ought to be seen and was in fact seen by contemporaries. It was Salt's personal response to

Saltaire United Reformed Church, the location of the Salt Family Mausoleum.

the pressures urging peace and stability between Capital and Labour which emerged in the aftermath of Chartism.

Sir Titus Salt died on December 29 1876. The legend of his charity, compassion and business acumen had spread far and wide long before his death, travelling throughout the United Kingdom and far beyond its shores, where his name and the manufactured cloth associated with it was known and appreciated. He was known to Queen Victoria, who had sent alpaca fleeces from Windsor and was delighted with their transformation at Salts Mill into lustrous cloth. He had been presented with the Grand Medal of the Legion of Honour in 1856 for the excellence of his products and was a great pioneer in a new era for the United Kingdom. His funeral procession from Crow Nest to his final resting place in the mausoleum at Saltaire's Congregational Church was to take many hours; most mills were silent and many workers lined the streets to see his funeral carriage passing, the numbers assembling estimated at around 100,000. Would his legacy continue through his sons, who would hold as their inheritance his creation of a grand, vertical, worsted mill and the industrial community of Saltaire?

2

The Salt Descendants and the Stead Family 1876–1893

Given his great success in the worsted textile trade, Sir Titus Salt left a legacy that was not merely that of finance, belongings and land to his wife and children, but one that also involved his vision of sustaining good relations between 'capital' and 'labour' through the worsted business at Salts Mill and the industrial community of Saltaire. This came through all the social, welfare and educational facilities that he had established. Through a number of key provisions in his will, however, he inadvertently left some significant financial problems for his executors. What were these provisions and how did these and the other prevailing trade circumstances affect his legatees in continuing his legacy? In Victorian times, with or without any difficult testator provisions, this depended almost entirely on the circumstances and personalities of the testator's male descendants – in this case, the five 'Salt' sons. Two sons had left the family business before Sir Titus died, and they were never to return and direct the future of Salts Mill and the industrial community of Saltaire that he had founded.

William Henry (1831–1892), the oldest son, married Emma Dove Octavianna Harris in 1855. Emma was the daughter of J. Dove Harris, Mayor of Leicester in 1856 and MP for Leicester from 1865 to 1874. In 1865, William Henry bought an estate of 250 acres at Mapplewell in Leicestershire and a handsome mansion. He began to devote himself to the life of a country gentleman and at his death was noted, in a short obituary recorded in the local Leicestershire press, to have lived a quiet and unostentatious life with little taste for the practical life of business. He is buried in a modest grave outside the church of Woodhouse Eaves near Mapplewell, and had chosen to avoid the grandeur of the family mausoleum at Saltaire in death. His father had had time to accept that his oldest son, who would inherit the title, was not very interested in the business or the industrial community he had founded. It is a matter of speculation as to how Sir Titus and his wife Caroline may have felt about this. It is clear from the last will of Sir Titus Salt (drawn up in 1871) that, prior to his death, he had advanced the sum of £32,000 to William Henry, as part of his final legacy to him of £100,000. On his death he additionally bequeathed a further £100,000 to be held in trust for the upkeep of the baronetcy, which would seem to indicate that he had no specific concerns about William Henry's choice as future baronet to become a gentleman landholder some distance from Saltaire.

Sir Titus's fourth son, Herbert (1840–1912), had, at the age of twenty, also decided on a very different path to that of working for the family firm. The 1861 census for Methley Hall, near Leeds, records Herbert as being a farmer with 316 acres and six men in his employ. By 1869 he appears in electoral and other records as a farmer of 450 acres, employing thirteen men and four boys at Thornes Farm, Beaulieu, Hampshire. Subsequently in 1872 he is recorded as 'managing the farm well and keeping the premises in much better order than any other tenant'. By August 1873, according to newspapers at the time, he was 'quitting Thornes Farm'. It has not been possible to track where Herbert was living between August 1873 and 1875, but he had evidently returned to North Yorkshire by 1876 (the year of his father's death) when he is recorded as living at Carla Beck, near Skipton, and farming variously at Bell Busk, Coniston Cold, Gargrave and Kirby – most probably, once again, as a tenant farmer. From the will of Sir Titus, it is clear that Herbert also had been advanced a sum of £10,000 during the life of his father. It is not possible to judge when the advance was provided and for what it was used.

Carla Beck House today.

All records for his farming ventures indicate that he was a 'tenant farmer' rather than a landowning farmer.

Press records indicate that Herbert left Yorkshire in 1884, and there are a few records of his travels to New York and Boston between 1884 and 1886. Somewhat surprisingly, from 1887 he is recorded on electoral rolls and successive census as living initially in Brixton and later at Wandsworth, Clapham, London – then an area recorded as being deprived by Charles Booth on his London Poverty Map 1898. He married Elizabeth Farrell in the parish church of St. Bride in January 1889. Elizabeth died in 1898 and her shares in the Great Western Railway, worth £3,880, were transferred to him. Elizabeth asked that her infant children be educated in the Roman Catholic Church. He married a widow, Margaret Ann de Lacey, in 1899, and died in Clapham at the age of seventy-two years. Throughout his time in London he is recorded on census returns and electoral rolls as being of independent means, but the source of these 'means' is not a matter of record. It is impossible to assess whether Sir Titus had a different perception of Herbert compared to his other sons, and he was never recorded as a partner in Salt's business, but he was treated even-handedly in the matter of the direct bequests made by Sir Titus Salt in his will.

Sir Titus Salt provided for direct legacies to the dowager Lady Caroline Salt (who was also left the house, grounds and all belongings at Crow Nest), William Henry and Herbert and some small financial sums to servants and others. There were legacies of £80,000 to each of his daughters (Amelia, Helen and Ada), £30,000 for the sick and poor in Shipley, and the £100,000 for the upkeep of the Baronetcy, all of which were to be held in trust and the will directed that these be invested by his trustees. The total amount of these legacies was £530,000, of which £370,000 was placed in trust. In addition, annuities were granted to Lady Salt of £5,000 p.a. and to six nieces of £100 p.a. (two other family members granted annuities had died before Sir Titus and these could not be applied), making the total of annuities bequeathed £5,600 p.a.

The large legacies to his daughters and the balance of the equal legacies of £100,000 for William Henry and Herbert were due to be paid in fifteen instalments, starting five years after his death (i.e., in 1881). When probate was granted in 1877, however, the direct legacies (those to be held in trust and the remaining annuities) far outweighed the then valuation of Sir Titus Salt's estate at under £400,000, and most of Sir Titus's estate would probably have been tied up in his business assets.

No direct legacy was provided for George, Edward and Titus Salt Junior, who were to receive an equal share of the residuary personal estate of Sir Titus Salt after settling all expenses and providing for legacies and annuities. Here the will directed that monies to cover all the legacies and annuities that were not required to be paid out immediately should be invested in appropriate company or government stocks, shares, debentures and bonds.

Unfortunately, as it was later to prove, the decision was taken to place all these monies in shares and debentures of the 'Sir Titus Salt (Bart) Sons & Company' – the family company owning and managing Salts Mill. The will having required that the capital from the legacies placed in trust be paid to the legatees in annual instalments over a period of fifteen years also directed that this capital should earn interest of 5 per cent per annum in the interim. It was probably impossible for this to be achieved without liquidating the assets of the company, meaning the loss of family control of the business; the decision to retain all of it in the company was no doubt made to honour its founder. The executors to the will were named as George, Edward and Titus Junior – so long as they remained involved in the business for a period of fifteen years (later changed by the first codicil to Sir Titus Salt's will to twenty years) after their father's death. Together with William Henry Salt and a friend, Alfred Allott, they were appointed as the five trustees of the Salt estate.

What is clear (with hindsight) is that, while both William Henry and Herbert had shown little or no interest in the management of Salts Mill and Saltaire and that any interest in this was to lie in the hands of George, Edward and Titus Salt

Extract from the Second Codicil to Sir Titus Salt's will, revoking Alfred Allott. (© Saltaire Archive)

Junior, their father had not provided for these three sons with direct financial legacies in the same way. While his will of 1871 clearly intends to provide for direct legacies of £100,000 for each son, should they have ceased to be involved in the business, by remaining involved these three sons inherited his residuary personal estate only after settling all expenses and providing for direct legacies and annuities. This was to create a financial burden for them in trying to ensure that the business could produce sufficient profit to ensure that their father's commitments to the will's beneficiaries were met, as well as to meet their own and the business's investment needs. In addition, there was to be a dramatic twist involved in the story of the trustee appointed in the will who was a close personal friend of Sir Titus Salt, one Alfred Allott of Sheffield. This friend's position as trustee was removed by a second codicil to the will in July 1876, not long before the great man died.

In the appointment and removal of Alfred Allott lies an interesting story that would also influence the future of the Salt family management of the mill to a significant degree in the last quarter of the nineteenth century. Richard Frost (2005), in writing a brief history of the Sheffield accountancy firm of Hawsons, provides some information about this friend of Sir Titus held in such regard that he was named, in 1871, as the only non-family member to be an executor responsible for carrying out his final wishes. Alfred Allott was born in Walsall in 1842 and his father, Robert, was a Congregational minister. He was educated at Silcoates school near Wakefield and, after a successful early career at the Sheffield and Hallamshire Bank, where he was highly regarded, he left the bank in 1854 to establish a new firm of public accountants. Here Allott prospered and his business interests grew. He was auditor for the Midland Railway, well known and trusted in the district and developing his broader business interests at a time of rapid industrial growth in Sheffield. In 1867 he entered into partnership with John Crossley, then an MP for Halifax, to acquire Brightside Colliery and in 1873 he bought his partner out. He went on to acquire the Ruby Mine in Cornwall, assumed control of Renshaw Ironworks in Elkington and the Newbridge Iron Ore company, and was persuaded to make an investment in America that was to prove to be his downfall.

He was involved with Sir Titus Salt for a number of years, no doubt partly because both his family and Salt's were prominent Congregationalists and possibly through the connection with John Crossley of Halifax, whose niece Catherine would marry Titus Salt Junior. Whatever the nature of the religious and family connections between Allott and Sir Titus, their bond was strong enough for Allott to approach him when his business ventures began to falter. Sheffield's period of prosperity was ending in 1873 and a number of companies were 'going under', but it was in this year that Allott had been persuaded by a London financier and an American general to purchase 40,000 acres of lands at Smith's Crossroads in Tennessee and Georgia that were believed to possess very valuable beds of iron and coal. Due to delays in building the Cincinnati southern railroad to exploit these deposits, Allott's investment in this land and mineral rights grew from £10,000 to £132,000 of largely borrowed money. By 1876 many of Allott's other investments in the United Kingdom had fallen in value and many of his small creditors were pressing for payment.

He turned to Sir Titus Salt for help, offering the land in the USA as 'security' in return for providing a loan, and in 1875 Allott, Titus Junior and Herbert Crossley went to visit the site and obtain reports from competent mineral engineers as to its value. The reports must have been satisfactory and a loan of £63,000 was made in 1876. Within a year, Allott was sued for bankruptcy by his creditors and the result of the hearing was that all his assets were liquidated, by arrangement, not in bankruptcy, by 1877. This was undoubtedly the reason for Allott being removed as a trustee from the will by the second codicil to the will of 1876, when Sir Titus must have been aware of Allott's impending demise. The loan could not be repaid and the Salt Company had the land, in what was later to be named Dayton, Tennessee, by default. This left the executors to the will and the business partners in Salts with another problem. Should they sell the land or regard it as an investment? The decision was eventually made to develop the site for the mining of coal and iron ore and the production of pig iron. Just how much Sir Titus himself was actually involved in the early detail of the affair is unknown. He was very near to the end of his life and it is possible that the remaining partners in the business – Charles Stead, William Stead, George, Edward and Titus Salt Junior – were the ones later persuaded of the value of the investment. It was to be Charles Stead and Titus Junior who were to travel many times to America to further develop the potential of this investment, rather than Titus Junior's older siblings George and Edward.

Before an exploration is made about the eventual outcomes for the Salt family business and the future ownership and management of Salts Mill, it is worth placing briefly on record what is known about George, Edward and Titus Salt Junior; Charles Stead and his sons; and William and, much later, Charles Frederick, who were to become directors in the company in the future.

Little is known or publicly recorded about George and Edward Salt, despite their equal legal partnership in the business at Salts Mill. Edward (1837–1903) married Mary Jane Susan Elgood in 1861 and had a grand house built at Ferniehurst, Baildon, in 1862. The Ferniehurst estate was extensive and its 'towered' mansion had seven bedrooms on the first floor and five bedrooms on the second floor – these five were probably for servants. It had a generous ground floor with a library and business room, in addition to rooms for general family purposes. The grounds had outbuildings that included a lodge, stabling, a model farm and six 'closes' of land, with woodlands. Edward had leanings towards the traditional life of 'the gentry' and was one of only two of the five brothers (William Henry being the other) to take out a gun license and engage in field sports. He was an enthusiastic cultivator of orchids in his extensive conservatories at Ferniehurst. He became

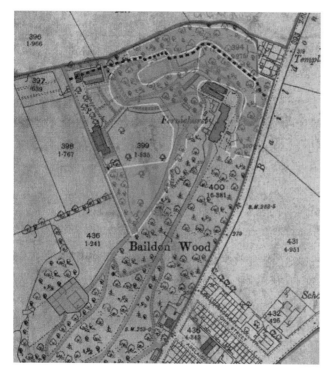

The Ferniehurst estate, Baildon.

a member of the Conservative party in the 1860s and was to join his wife as a member of the Church of England. After his first wife died in 1870, he married Sarah Amelia Rouse, the granddaughter of William Rouse, who had provided Sir Titus with vital early training in the worsted business.

George (1833–1913) gained a local reputation for being a clever amateur mechanical engineer and did not marry until he was fifty-two years old. There are reports of his constructing a handsome road carriage, driven by a horizontal steam engine and boiler and able to carry ten passengers in 1870. His most extravagant purchase seems to have been a luxury yacht that he kept moored at Scarborough and, for many years, he lived the conventional life of the wealthy bachelor. He organized the annual Bradford Bachelors' Ball and caused tongues to wag when he was accompanied by theatre actresses at the Royal Yorkshire Exhibition in 1887. He married Jenny Louise Fresco in 1885 and neither he nor Edward were to have children.

George was to cease to be a partner in the family business in 1881 when the partners decided to re-construct the company and adopt limited company status. At this point George had agreed to payment of £11,409 over two years, in exchange for the transfer of his shares in the business to Edward and Titus Junior. Edward remained working as a partner 'owning and managing' Salts Mill until the Salt connection was severed in 1892/3. Neither before 1881 nor after that point in time did either of these older siblings of Titus Junior appear to have become involved greatly in the kind of public life led by their father or in the social or other affairs of Saltaire Village. Their personal roles in the management of Salts Mill don't appear to have been publicly perceived as leading the direction and development of the worsted textile business built up by their father, although it is quite possible that they were heavily engaged in the day-to-day running of the factory. George's interest in and skills with mechanical engineering must have been of value in a business reliant on huge amounts of machinery until he ceased to be a partner in 1881. Edward continued as a director of the company until its final demise.

The two men who were to play the largest publicly recorded part in managing the mill and, to a very large degree, in sustaining the wider legacy of Sir Titus Salt through their efforts made in public life were Titus Salt Junior (1843–1887) and Charles Stead (1823–1902).

Charles Stead had worked for Sir Titus in his Bradford mills and, along with William Evans Glyde, was appointed as one of two junior partners at Salts Mill in 1854. It is worth remembering that Titus Junior was aged ten years when Salts Mill was opened in 1853, whereas Charles Stead was thirty years of

Titus Salt Junior.
(© Saltaire Archive)

Catherine (Crossley) Salt.
(© Saltaire Archive)

age at that point and would have had considerable managerial experience in worsted textile manufacture. He was recorded as 'Mr. Salts' manager' as early as 1853. There is little recorded or preserved in detail about Titus Junior's direct involvement in the management of the business at Salts Mill and, where this is available, it is usually couched in terms of his being 'the only son of Sir Titus to inherit his business acumen (sic) and to take on a similar role in political and public life'. As R. Lee Van den Dael and R. David Beale note in their book *Milner Field* of 2011:

> By the age of 22 he [Titus Junior] had become involved in the family textile manufacturing business and his ready grasp of the principles and strategies necessary to succeed astounded those who observed him.

Titus Junior had made a 'dynastic marriage' to Catherine Crossley, daughter of a wealthy Halifax carpet manufacturer, Joseph Crossley, in 1866. They honeymooned in Venice and Switzerland and initially lived at Baildon Lodge, where an 1866 bill for their furniture is recorded as amounting to £4,000.

Within a year of their marriage Titus Junior purchased 161 acres of rich arable pasture, together with an old mansion house, two farm houses, a coach house and stables, collectively known known as the Milner Field estate. Situated on a hill above Saltaire in Gilstead, Bingley, it was a short carriage ride's distance from the mill. There is a recorded transaction of purchase for the land at Milner Field for £21,000 by the Salt Family in 1869, but this was not to be the extent of the money lavished on this new home. The old mansion was pulled down and Titus Junior commissioned a little-known architect, Thomas Harris, to design a grand new house. This took two years to build and was completely different in design to the Italianate style of Salts Mill and Saltaire. It is thought that the extensive new house was completed at a cost of £40,000, and that its interior was as grand as its exterior. Titus Junior also had extensive conservatories constructed, large landscaped and kitchen gardens, a model farm and a boating lake that was well stocked with trout. The carriageway alone cost £6,000. How such a young man, even one with a wealthy father, could have afforded the overall expenditure undertaken remains a mystery. It is possible that some of the money available to the young couple was provided by the equally wealthy Crossley family.

Milner Field, Architect's drawing. (© Saltaire Archive)

Titus Junior had moved with his parents and siblings from the poorest areas of Bradford very shortly after his birth in 1843 and had attended Mill Hill school for a much shorter time than his older brothers; consequently, his experiences were largely that of luxurious surroundings and probably guided to a greater extent by his mother's religious devotion. He entered the family firm in 1863 aged twenty years and, within two years of his involvement, his eldest brother, William Henry, and the only other non-family partner, William Evans Glyde, had retired from the business.

As early as July 1865, Titus Junior is reported as subscribing £26 to the foundation and building of the Northern Counties Asylum, and he rapidly extended his political, charitable and public profile. In 1867 he was appointed as a trustee of the Bradford Infirmary and Dispensary fund; had donated money for the proposed Bradford Fever Hospital; was appointed a vice president of the Royal Albert Asylum for Idiots and Imbeciles of the Northern Counties; contributed money to a new lifeboat station in Milford Haven; and was recorded as attending a number of meetings of the local Liberal Party. His progress in philanthropic, political and public life continued and, in the years 1867 to 1876, alongside a number of activities in Saltaire, he had become the vice president for the Bradford Working Men's Club; supported the building of a new Congregational church in Bowling; laid the foundation stone for a new Congregational church in Ilkley; become vice president for the Workmen's International Exhibition; been appointed treasurer for Airedale College; and been selected as one of a group of gentlemen to represent Bradford at the London Conference of the Liberation Society. He had also become father to his first son, Gordon Salt.

His father must have been sure that Titus Junior would be the most likely to emulate his own deeply held sense of social responsibility, as one of the first provisions in his will of 1871 was that the almshouses, the dispensary, the precinct for the Congregational church and the family mausoleum in Saltaire be directed for the use by his wife and Titus Junior for their sole benefit as joint tenants. Titus Junior went on to be elected as a Bradford town councillor (East Ward); to chair the Shipley Schools Board for many years; to create a governing body and preside over this for the Saltaire schools; to attend a large amount of Bradford Council meetings; to support passionately the incorporation of Shipley into Bradford (which was so largely opposed that it was dropped and not achieved until 1974); and to initiate many social events and some technical experiments in Saltaire Village, of which the best known are probably the establishment of an annual conversazione and 'magic lantern' shows for the public within the Saltaire Club and Institute (later known as Victoria Hall) and his experiments with telephone communications between Milner Field and Salts Mill and, later, between Milner Field and Halifax.

Charles Stead, who was recorded in the 1851 census as a manager in a worsted mill, living in Horton, Bradford, with his family and one servant, had moved with his family to Baildon by 1856 and had had a significant head-start

A ball at Victoria Hall. (© Saltaire Archive)

on Titus Junior in respect of philanthropic, political and public life. He was a fellow Congregationalist and supporter of Liberal political views, and his public life was notably more directly involved in both business networks, where he became chair and later the president of the Bradford Chamber of Commerce, and in another local business venture, the Bradford Exchange Company. He was, for many years, the chair of the Shipley Local Board (the then local authority for the area in which Saltaire is situated) and presented the arguments on a number of occasions for additional investment in Shipley's water supplies and road improvements. He was a notable presence in delegations to the Foreign Office in respect of treaties between the United Kingdom and North American states and in meeting the Home Secretary in 1875 as a representative of the Salt Company, arguing against some of the provisions in a new Factory Act that would raise the age of child workers from eight years to ten years. It is particularly notable that it was Charles who received a deputation of workers from Salts Mill in 1876 after a walkout over pay; he was instrumental in locking the workers out and holding firm against their demand for better pay, rather than George or Edward Salt. From newspaper archives of the time between 1877 and 1892, Charles is the person most frequently named as dealing with events occurring at the mill.

Despite chairing the Shipley Local Board for many years, he supported Titus Junior in arguing for incorporation with Bradford and presided over a large meeting of ratepayers in 1880, but could not achieve their agreement to the proposal. He was much less involved in the social life of Saltaire, attending important functions such as the annual conversazione, with his oldest son, William. William was eight years younger than Titus Junior, and he also became a director of the Salt business when the opportunity arose. By 1881 Charles was recorded by the census as being a worsted manufacturer living with his family, a governess, a cook, and four maids at the Knoll, a large house in Baildon. He had progressed in both financial and social terms, and played a significant role in changes that were made in the structure of the company, Sir Titus Salt (Bart) Sons & Company, in 1881.

By 1881 an international economic crisis was well established and what was known as 'the panic of 1873' (which had caused the collapse on the Vienna Stock Exchange, driven by booming stock prices in central Europe) had reached fever pitch. It culminated in a 'market panic' for a relatively short period but was of such significance that it was to spark a major economic downturn. One of the many causes of this panic was the construction of the Suez Canal, which opened in 1869, because goods from the Far East had been travelling in sailing vessels around the Cape of Good Hope and were stored in British warehouses. These sailing vessels were not adaptable for use in negotiating the new canal and British trade suffered.

Indeed the 'panic' of 1873 and a multiplicity of other factors prevailing in Europe and the United States of America (just recovering from the Civil War) was to result in the 'long depression' thought by many historians and analysts to be the first truly international economic crisis. This is seen by many to prevail from 1873 to 1896 – overlapping a little with the end of Sir Titus Salt's leadership of the business at Salts Mill and covering the whole of the period when three Salt siblings and the Stead family members were directing the business. The 'long depression' is thought to have hit the United Kingdom the hardest, and during this period the United Kingdom lost some of its large industrial lead over the economies of continental Europe. It also led to bankruptcies, escalating unemployment, a halt in public works and a major trade slump that lasted until 1897.

The United States of America were also badly affected and the primary cause of price depression in North America was the tight monetary policy that the USA adopted in its efforts to get back to the gold standard after the Civil War. The government took money out of circulation to achieve this goal, meaning there was less available to facilitate trade. It is this context that was to have a bearing on the land acquired through the default of a debt (originally known as Crossroads) and the Salt company investments in iron ore and coal deposits in what became Dayton, Tennessee. The American business sectors that experienced the most severe declines in output were manufacturing, construction and railroads, and the expansion of railroads came to a dramatic end in 1873. Between 1873 and 1878 the total railroad mileage in the United States barely increased at all, which was in all probability a major factor in Allott's inability to realise his investment there. David Ames Wells (1890) wrote of changes within the period of depression that included the reduction in warehousing and inventories, the elimination of 'middlemen', increased importance for economies of scale and the decline of craftsmen. He also noted that these changes created great disturbances in old methods, and entailed losses of capital and changes in occupation for many individuals.

How this first truly international economic depression affected the Salt business in terms of trade and profitability cannot be determined, due to a lack of preserved and complete accounting records for the period. Whatever effect it had, the actual payment of interest on the legacies and annuities of Sir Titus's will between 1876 and 1881 is not known, but the five years required by the will prior to administering the remaining provisions for his legatees had elapsed. So at the very least it had by then become important to adopt a new, incorporated/limited company structure by which the settled funds under the will could be invested in shares and debentures of the limited company, with each family legatee receiving shares and debentures to the nominal value of their legacies. 1881 was the first year when the instalments of the legacies were due for payment. Up to this point, if the family legatees had received interest of 5 per cent each year on the outstanding unpaid balance of their legacies, the total sum to be paid out would have amounted to £24,900 per annum. If such an annual sum had been paid, this in itself would have been a big drain on the estate's resources. The newly incorporated company, formed in 1881, had its directors listed as Edward Salt, Titus Salt Junior, Charles Stead and William Stead, Charles's oldest son. It is of note that Salts Mill and the business of worsted manufacture it carried out was, from this date, under the equal leadership of the two remaining Salt siblings and two members of the Stead family, and that the company had become Sir Titus Salt (Bart) Sons & Company Ltd.

An 1883 'Memorandum and Suggestions' document for Sir Titus Salt (Bart) Sons & Co. Ltd reveals that, only two years after incorporation as a limited company, the company was not performing well. It had a deficit in its balance sheet of £26,163 and an estimated £150,000 at risk in the Dayton investment. The document notes that

> the residuary legatees under the will of the late Sir Titus, who were deputed by him to succeed in the business and on whom he intended to confer a preponderating share of his accumulated wealth, have hitherto received little more than a legacy of anxious care.

At this point the family were agreeing to receive less than what was due to them from their holdings in the new firm, in an attempt to keep the company afloat. Newspaper and other archive sources have been explored to seek clues

The Prince of Wales planting a commemorative tree in Saltaire Park, 1882. (© Saltaire Archive)

as to whether any changes in the public and social activities of Titus Salt Junior and Charles Stead might indicate a growing concern for the business. In 1882 a prestigious event was arranged as 'the result of an interview between Titus Salt Junior and the Prince of Wales at Marlborough House', as the *Yorkshire Post* for 25 April announced. The report included detail of a programme of arrangements for a royal visit to Bradford and Saltaire where the Prince of Wales and his consort were to arrive, as guests of Titus Salt, on June 22 and attend a banquet he had arranged that day, travelling to Bradford the next morning to open the new technical school for the town. Charles Stead also played a role in making arrangements for this royal visit and is recorded as being introduced to the royal couple.

Neither Titus Junior nor Charles appear to have 'slowed down' or ceased their activities in public life, despite the financial and business climate; for example, in November 1883, Charles was elected on to a committee to take forward the proposal for Bradford to have a central railway station and he was also elected to sit on the council of the Bradford Chamber of Commerce in December of that year. Notably, also in December 1883, it again was Charles who met a deputation of workers – a meeting that resulted in an agreement to end another strike by operatives at Salts Mill. Similarly, Titus Junior was among a group of gentlemen meeting with the Mayor of Bradford to discuss opening a charitable fund for the needy of the city. In July 1883, however, it is recorded that he reported his intention to resign from the Shipley Schools Board because the board had rejected a policy of exclusion for 'half timers' from Shipley Schools (the only education available for children working in the mill for half of each day) and he did resign from the Shipley Schools Board in September 1883. These and other clues indicate that the trade depression was forcing a 'hard attitude' on the part of employers.

What is of great interest in this account is that, in February 1883, the Dayton Coal & Iron Company was incorporated as a limited company in England and some 23,845 acres of land inclusive of mineral rights and a further 3,844 acres of mineral rights were transferred from Sir Titus Salt (Bart), Sons & Co. Ltd to the Dayton Coal & Iron Company. In May 1833 the *New York Times* reported that Titus Junior and Charles (both incorrectly reported as being titled) were in the United States of America to visit Dayton, Tennessee, 'with a Mr Howson and two other gentlemen, who are among (sic) the leading manufacturers of iron and steel' and who are in America 'to visit the Dayton coal mine, 25 miles from Chattanooga, to make extensive mineral explorations, examine its water power and resources generally ... with a view to making large investments in the manufacture of iron, steel and cotton'.

Even with the uncertain accuracy of journalistic understanding, it is made clear by this report that the decision had been made by the Salt Company to proceed to invest in the land originally gained by default due to Allott's debt to the Salt family in 1876. It is also quite possible that the decision had been made in order to support the Salt's UK textile business through diversification and investment in America in new products. By 17 May 1884, Titus, Charles and a William Donaldson

The blast furnaces at Dayton.

of Glasgow arrived in Dayton and, according to the American press, announced that they would erect two large blast furnaces with a capacity of 250 tons of iron ore per day and invest $500,000 dollars. Notes taken by Saltaire historians Shaw and King, during a visit to Dayton in 2012, record that in September 1884 the Dayton Company had been reorganised under the same name with a capital of £1,200,000 and that the firm of James Watson & Co. Ltd, Glasgow, had been brought in as partners/fellow investors with six of their own directors.

Prior to this, Shaw and King's notes extracted from Rhea County Historical Archive, record that on 6 January 1882 rails arrived from England. 8 miles of railroad track were eventually laid and the Richland Mine was opened. In 1884, the foundations for the blast furnaces were laid, three fire brick kilns were built, 200 coke ovens were constructed for the mine and –

200 homes were constructed as were a managers' house (costing $15,000).
A company store, a schoolhouse and other social amenities were established.

There is a sense of déjà vu between these facts and the planning and building of Saltaire (less than thirty years earlier), and what had been a very small American 'hamlet' certainly grew into a sizeable town over a very short period. Records about the level of capital investment in the 'Dayton venture' from the Salt company in comparison to the Watson company from Glasgow in 1884 are not available but, as noted earlier in the chapter, £150,000 was seen to be at risk in 1881 for the Salt company.

In Saltaire, Shipley and Bradford, life seems to have continued for Charles Stead, and to some degree for Titus Salt Junior, in the pattern that had been established in the 1870s. Charles continued as a magistrate for the West Yorkshire Courts; remained involved in the Shipley Local Board, the Bradford Chamber of Commerce and the Bradford Exchange Company; was newly listed as being a

Dayton, Tennessee, c. 1910.

director of the Yorkshire Board of the Economic Fire Office Ltd; and in 1886 was again recorded as Chairman of Sir Titus Salt (Bart) Sons & Co. Ltd. He continued to be involved in Liberal Party political affairs and was often noted as holding positions or chairing committees, some of which were established specifically to support ventures initiated by Titus Junior.

There is evidence of some change in Titus Junior's public activities after 1883 and he is more frequently recorded as being involved in events with a social or charitable nature than in some form of formal public life. For example, he attended the sports day for Bradford Boys' Grammar School in the Saltaire Park, attended a meeting of Bradford's Lord Mayor to consider establishing a relief fund for the poor in Bradford and presided over an operetta performance in Victoria Hall that 1,200 people attended between 1883 and 1886. As noted earlier he had resigned from the Shipley School Board in September 1883. In November 1884, however, Titus wrote to the Shipley School Board to complain that 'the affairs of the board are being inefficiently managed' – suggesting that his resignation had not been a 'happy' event for him and that he was still intensely interested in the affairs of Shipley's school management. He remained as chairman of the Saltaire school governors and was appointed as a governor for Bradford Girls' Grammar School. He attended some important Liberal Party meetings but declined an invitation from the Shipley Liberal Party to be their candidate for the parliamentary division in 1885.

Salts high school building. (© Saltaire Archive)

It is known that Titus Junior was diagnosed with a heart condition around two years before his death in 1887 and his knowledge of this condition may have been a reason to change the focus of his public involvements. Nevertheless, both Titus Junior and Charles are recorded as making a number of visits to America in connection with the Dayton company during these years. In July 1884 Titus Junior is recorded as being in Dayton, without Charles, to 'close a contract to establish and iron ore plant costing £100,000'. These journeys, the worldwide trade depression and his efforts to fulfil his father's wishes must surely have been a heavy burden that affected his health. Despite these circumstances, or perhaps because of them, in October 1886 Titus proposed to 'raise a memorial to his father' by building a new 'School of Art and Science to be erected on land behind Victoria Hall', simultaneously proposing an 'International Exhibition' at Saltaire to celebrate Queen Victoria's Jubilee in 1887. He announced that the new building would be designed by Messrs Lockwood and Mawson in the Italianate style; it would house fourteen large classrooms, have a central museum hall – with the dimensions 70 feet by 40 feet – and a gallery running all around this. The building costs of around £12,000 would be defrayed by public subscriptions and income to be generated from visits to the planned exhibition, which was to run from May to October 1887.

Titus Junior, as chair of the Jubilee Exhibition Committee (established by the Saltaire schools governors), met aldermen and gentlemen in the Mayor's parlour at Bradford Town Hall in January 1887 to further the plans for the exhibition and for the reception of the royal visitors, who were to attend and perform the inauguration ceremony for the new school and the opening ceremony for 'the Royal Yorkshire Jubilee Exhibition'. The royal visitors were Queen Victoria's youngest daughter Beatrice and her husband, Prince Henry of Battenberg, and they were to be guests of Titus Junior and his wife Catherine at Milner Field, arriving on 5 May 1887. The building of the extensive new school was achieved by April 1887 and a dinner was given in the dining room of the 'Exhibition Building' by Mr Titus Salt; 'there was a large company, consisting to a considerable extent of press representatives'. Charles meanwhile established a committee from within the membership of the Shipley Local Board to agree the steps to decorate the royal route for the occasion, for which the committee are to 'solicit subscriptions' (Charles and Titus Junior guaranteed £50 each to the fund). At its closure, the exhibition recorded that 823,133 visitors had attended but it was noted by the local press as 'not being as successful as had been hoped'.

In the event, as reported by the *Yorkshire Gazette* in November 1888, the financial result of 'the Saltaire Exhibition' was that the new Science and Art School, built at a cost of £11,000 as a memorial to the late Sir Titus Salt, had only attracted subscriptions of £2,580 and that the 'royal exhibition' had made a loss of £4,000 and that these together left a debt of (*sic.*) £12,000 on the building. A letter from William Fry to the Salt family in March 1889 confirms that this had left the accounts of the Saltaire schools 'in a very bad state'. Titus Junior must have been very aware of the large financial problem that his efforts were going to create for the Saltaire schools, in which he had invested so much of his time and energy. On 18 November 1887 he was due to preside over a meeting of the 'executive committee for the exhibition' at Salts Mill that afternoon but, at 12 noon, 'due to feeling indisposed', he sent his apologies for not being able to attend and he left for home (Milner Field) at 12.30 p.m. where he apparently had his lunch, took a walk around the grounds and returned to the house at 3.45 p.m. According to reports at the time, he entered the smoking room to rest himself but was 'taken immediately worse' and his butler summoned Mrs Salt, but when she reached her husband's side he was unconscious and very shortly afterwards died.

The majority of the obituaries for Titus Salt Junior were long and praiseworthy. He was noted as having 'inherited much of that business acumen and enterprise so conspicuously displayed by his father' and perceived as having pursued his father's goals to promote the social and intellectual wellbeing of the people 'among whom he lived' (a somewhat generous view, given the luxury of his home compared to Saltaire residents of the time). He was given particular praise for his establishment of a fine group of Saltaire schools in Albert Road and for supporting the innovative methods of education established there. Although his wife Catherine Salt was also heavily involved in the schools, she is not mentioned in the local obituaries.

It is beyond the scope of this book to attempt an analysis of the character of Titus – although this is an area ripe for further research – but there are some strong indications from public and press records that suggest he was completely committed to his father's moral, political and socially responsible approach to life. Whether he had perhaps not acquired quite the expertise and business acumen in managing Salts Mill or whether global economic depression, coupled with the burden of administering the difficult provisions of his father's will, were the main causes of the financial problems left behind for the Salt company is for others to judge. His funeral took place on 23 November and his remains were interred in the family mausoleum at Saltaire United Reformed Church.

The company of Sir Titus Salt (Bart) & Sons & Co. Ltd would continue for another four years with Edward Salt as the only remaining Salt family member acting as a director. Charles Stead and his oldest son William also continued as directors, quite possibly appointing others during these years (as indicated when the company records chart its considerations of the advisability of winding up the company in September of 1892). Charles Stead's pattern of public and charitable engagements hardly change after Titus Junior's death. He was elected to the Liberal Hundred in November 1887, became a director of the Liberal Club Buildings Company in 1888, and was involved in securing a representative display of the district's manufacture for the Paris Exhibition of 1889. In November 1888 he was selected as the only candidate for the county council elections, Shipley division,

The Salt Family Mausoleum.

and was duly elected in 1889, becoming an alderman for the county council. He continued with his commitments for the Bradford Exchange Company; was invited to sit on a new joint hospital board for Shipley and Windhill; attended meetings of the Salt School governors; and, from January 1891 to September 1892, the records of his public and political life continued to show a wide involvement in the district's affairs.

Charles also remained involved in the Dayton Coal & Iron Company, and records show that in July 1889 sixty new coke ovens were established for 'the Nelson mine'. In 1890 electric pumps and a steam conveyor were installed, and a worker's strike occurred in the same year. In 1892, however, James Watson & Co. Ltd 'took over' the Dayton company and the Salt company ceased to have any direct involvement. Charles, however, visited America a number of times during this period with his younger son, Charles Frederick Stead, who emigrated to America in 1890.

In December 1890, the American press reported that a group of British capitalists, listed as being Messrs R. and J. W. Pearson, Mr Taylor and Mr J. Baker of Saltaire, and Mr Charles Frederick Stead of New York, had purchased a large factory at Bridgeport, Connecticut, to start the manufacture of 'plush' (velvet). The reason given for this British investment in America is to avoid the 'prohibitory effects of the McKinley Tariff', and the report states that 100 families and 200 looms were to be brought to the site from England to commence this production. The Tariff Act of 1890, commonly called the McKinley Tariff, was an act of the United States Congress, framed by Representative William McKinley, that became law on 1 October 1890. The tariff raised the average duty on imports to almost 50 per cent.

The Warping Room at Salts Mill. (© Saltaire Archive)

1892 was the year that the Salt company were to acknowledge the company's severe financial problems publicly, and it is interesting to note that Charles Stead neither reduced his political and public involvements in the Bradford and Shipley districts, nor ceased to find other ways to secure the financial prospects for his younger son at least. He was aged sixty-nine in 1892 but his energy seemed undiminished.

The future for Salts, however, was looking very bleak. On 3 September 1892 a meeting of the directors and shareholders of Sir Titus Salt (Bart) Sons & Co. Ltd was held at the offices of the company in Saltaire to consider winding up the company. William Henry Salt had died in July that year and among those present at the meeting were Charles Stead (presiding as chair of the company), Edward Salt, George Salt, Herbert Salt, Sir Shirley Salt, Bart (oldest son of William Henry), Gordon Salt (Titus Salt Junior's oldest son), William Stead, Mr J. R. Armitage and Mr J. H. Wade. The meeting was held in private and, with the exception of Edward, all the remaining sons of Sir Titus Salt were present as major shareholders. A statement was issued to the press after this short meeting. The statement confirmed that it had been agreed that it was advisable to wind up the existing company and it explained that, since the issue of circulars summoning the meeting, Mr Charles Stead and Mr Joseph Croft (manager of the Bradford Banking Company) had been appointed to act as receivers, managers and liquidators and that, while possible schemes of reconstruction had been discussed, nothing definite had been agreed.

A document dated November 1892 does provide a 'Statement and Scheme for Reconstruction' and sets out the details of the proposed takeover of the mill and the village of Saltaire by a consortium of four local businessmen, who are listed as Isaac Smith, chairman of John Smith & Son Ltd, worsted spinners, Field Head Mills, Bradford; John Rhodes, colliery owner of Snydale Hall, Pontefract; John Maddocks, chairman of the Bradford Manufacturing Co. Ltd; and James Roberts, wool merchant, of Bradford.

The statement begins by giving an outline of the will of the late Sir Titus Salt; goes on to outline the capital structure of the company; describes the company's incorporation in 1881, its recent decision to apply for voluntary liquidation and its progress in doing so; and, finally, details the proposed takeover by the consortium under the heading 'Scheme of Reconstruction'.

Saltaire historian David King has made extensive notes on the main points in the statement, which help to clarify why little financial benefit accrued to his descendants. King notes that the 'A' mortgage debentures are limited to £200,000, of which £199,970 remained outstanding and held by the public, and that no part of the principal monies of the 'B' debentures had been paid off and only one half year's dividend on any of the shares in the company had been paid in the eleven years since incorporation. In addition, the 1892 document states that,

The £150,000 of 'C' debentures relating to the three daughters, Herbert Salt and the charity for the sick and poor of Saltaire is secured by certain properties in the vicinity of Saltaire which are held by the Trustees of the late Sir Titus. These properties include the Park and recreation ground and, in addition, about 250 acres of land, part of which is considered to be building land. The land irrespective of the Park was valued in 1881 at £67,488, but the valuer had now stated that such value, or anything approaching it, could not be achieved in the near future. Total income from the properties is stated to be about £1,000 pa, irrespective of the Park, which is a source of annual cost and not of revenue. An agreement of 1 July 1881 is mentioned whereby George Salt transferred his share in the residue of the will of Sir Titus to his two brothers, Edward and Titus Salt Junior in consideration of an annuity of £3,500 pa. No payment had been made to George under this agreement.

Sadly, the wishes of Sir Titus Salt with regard to his children had proved impossible to fulfil and the future of the company and the industrial community he had founded – Salts Mill and Saltaire – were to lie in the hands of others for the future. From 1893, just seventeen years after the death of Sir Titus Salt, the mill and the village no longer remained in the hands of the Salt family, and the financial wealth Sir Titus Salt had no doubt intended, through his last will, to confer for the benefit of his family would not have been possible to realise without the liquidation of the company shortly after his death.

The role of Charles Stead in the company over many years is worthy of a brief final comment. There is a sense in which he also held a deep commitment to acting in a socially responsible way and shared the political and religious views of the Salt family. During his time at Salts Mill he does appear to have led the decision making of the directors and he had the seniority and experience to influence greatly the three Salt sons who sustained some involvement in the company, perhaps encouraging some of the risk involved in the Dayton affair. One cannot help but wonder either whether he simply got used to holding positions of power and influence, and was reluctant to cease public activities for the sake of focusing more on the state of the business at Salts Mill.

There are a number of indications that this may have been the case and, although he left Yorkshire in 1893, moving into 'Red Barns', Victoria Road, Formby, Lancashire, he continued to chair the Bradford Exchange Company for some years and for a short time continued to preside over the Shipley Local Board. He rendered himself ineligible to continue as a county councillor (Shipley division) but, by 1894, he had become president of the Formby Ratepayers Association; gained a position on the Commission of the Peace for the County Palatine of Lancaster; and in 1895 was selected by Formby Liberals as a candidate for Lancashire County Council,

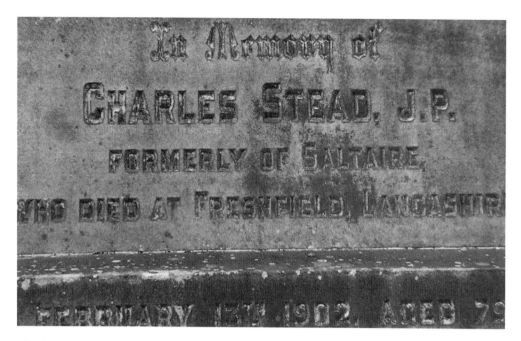

Charles Stead's headstone, Nab Wood Cemetery, Shipley.

Birkdale division (although not elected). He also continued to be involved in legal matters as a magistrate in Liverpool and gave evidence in June 1898 to a House of Commons Select Committee considering the law relating to burial grounds.

In February 1902, Charles died, aged seventy-nine years, at Red Barns after a short illness but, apart from attending Lady Caroline Salt's funeral in 1893, he does not appear to have sustained any contact with other members of the Salt Family.

3

Sir James Roberts and the Consortium Years 1893–1918

It is important for the reader to understand that, when the consortium of four Bradford businessmen purchased Salts Mill, the village of Saltaire and some holdings of land in and around the Shipley area making up the Salts estate, they did not change the name of the company founded by Sir Titus Salt. It remained known and registered as Sir Titus Salt (Bart) Sons & Co. Ltd throughout the consortium years, the years of sole ownership by James Roberts, and beyond his ownership for some time. The reader may recall from Chapter Two that the company had debts of £26,163 in 1885 and £150,000 at risk in the Dayton venture. Just what the level of debt was when the consortium offered to purchase the business is not clear but, between 1885 and 1892, both Edward Salt and Charles Stead had mortgaged their own residences to help shore up the company and by 1892 the banks had foreclosed on these mortgages, with the company entering voluntary liquidation.

The consortium's purchase price is not known but they rescued both the company and Saltaire in its entirety. The *Yorkshire Post*'s report of the purchase of the business and the estate reported that Isaac Smith was to be chairman of the new board; John Maddocks was appointed to manage the manufacturing and merchanting for the company; James Roberts was to be responsible for supplying the raw materials for manufacture; and John Rhodes (a wealthy colliery owner), the fourth director, was not recorded as having a specific role. The first part of this chapter will briefly explore the nature of all four men but the chapter will focus for the main part on James Roberts, who was to be awarded a baronetcy in 1910 and who can perhaps be regarded as the 'Second Lord of Saltaire'.

Isaac Smith (1832–1909) was born in Horton, Bradford, and was the third child of John Smith and Mary Jardine. His father John Smith owned Fieldhead Mills, a worsted spinning company, in Horton, Bradford, and is most likely to have been acquainted with Sir Titus Salt. In 1851 Isaac Smith is recorded on the census return as living with his parents in Horton and his occupation was given as book-keeper. By 1853 he had joined the family firm and in 1858 he married Sarah Holiday, who

Processing raw wool, Salts Mill, early 1900s. (© Saltaire Archive)

bore eleven children – five daughters and six sons. After a short time, the family were recorded as living at Chestnut Cottage and, later, Field House, Daisy Hill, Bradford – both sites are now integral to the Bradford Hospital Trust headquarters. There are early records of Isaac showing horses at agricultural shows and of his being elected to the Bradford Board of Guardians. His father, John, died in 1870 and he left the worsted spinning business to three of his sons: Benjamin, Henry and Isaac. Benjamin and Henry were to retire from the business during the 1880s and Isaac then ran the company with his sons. He had a remarkably similar pattern of life to Sir Titus Salt and Titus Salt Junior in that he had Liberal political affiliations, was a dissenter in religion and held a number of public offices.

Between 1874 and 1879, Isaac was elected unopposed as the Liberal councillor for Horton Ward (Bradford), and was recorded as laying the foundation stone for a new Baptist Church and a memorial stone for the Sunday school of Daisy Hill Methodist Church. He opened an art exhibition at Heaton Mechanics Institute and attended a public meeting in St George's Hall, Bradford, to protest against the war in Afghanistan, as did Titus Salt Junior. Between 1880 and 1887 Isaac was elected as alderman for Bradford Town Council; appointed as a magistrate for the Bradford Borough; was elected as the Mayor of Bradford; and was appointed as one of the first directors of the Bradford Central Railway (a company that folded just one year later). He continued with charitable works, and a newspaper report of 1883 records him as a teetotaller. As a worsted spinning company owner he experienced some worker strikes and was listed with Charles Stead as being on the Yorkshire Board of the Economic Fire Office (London). Most notably he was one of the five Bradford gentlemen appointed to a committee to promote the Yorkshire jubilee celebrations at Saltaire that were to prove to be so costly to the Saltaire schools governors and, perhaps, to be the final burden for Titus Salt Junior.

In January 1887, Isaac resigned as alderman for Bradford Council; reports indicate that he had been appointed as chairman of the Northern Counties Investment Trust in 1890 and that he prospered commercially, so much so that he was in a position to buy Allerton Mills in 1892 to add to his manufacturing outlets. There can be little doubt that he knew and was known by the Salt and Stead families, but what the relationships were can only be guessed at. The same *Yorkshire Post* article that reports on the consortium purchase of the Salt family business records a denial from the four men of some rumours abroad that the Salt family had been troublesome in the negotiations for the takeover of the business and the purchase of the estate. It also states that, at the time of purchase, the mill's workforce had reduced from 3,000 workers to 2,000 workers and concludes with a statement from the new board that it would invest in new engines and machinery and would produce a 'broad range of Bradford goods rather than specialising'.

Women and children, Salts Mill winding room, *c.* 1910. (© Saltaire Archive)

This indicates that investment in the Saltaire company had not been possible for some time and that the first thoughts of the new owners were to move away from specialised cloth.

It is without doubt that Isaac Smith was the most experienced of the four men in the manufacture of cloth and he was known to be connected to John Rhodes through the marriage of his only daughter, Esther Rhodes, to Isaac's second son Albert Smith in November 1885. The *Sheffield Independent* newspaper report of 15 November 1892 of the consortium's offer to purchase the Salt business and estate states that 'Mr John Rhodes is thought to be the financier of the scheme'. John Rhodes (1821–1913) was born in Bradford to Timothy and Sarah Rhodes, and his father was the publican at the Ship Inn, Well Street, Bradford. The June 1841 census shows John living with his parents when he was aged twenty and records his occupation as a wool stapler. He married Martha Esther Terry in 1845 and, at some point between the 1841 and 1851 census returns, he had changed occupation and become a coal merchant, having moved to Gildersome, 5 miles south west of Leeds, with his wife, two children and one servant. The indications that he was prospering continued and, by the 1861 census, he is recorded as a colliery proprietor, having moved house again to Tong, Bradford.

In 1867 John was reported to be the owner of Snydale Colliery, at Normanton, South Yorkshire, and in 1869 John and his family moved to the nearby Snydale Hall with six servants, further indicating that his early change of occupation had led to some significant commercial success. This success continued and he was able to persuade the trustees of Pontefract Park to allow his growing business a lease for fifty years for 'the getting of coal' under the park for the rent of £600 per annum. This colliery was to be named the Prince of Wales Colliery and, after two years' exploration and the sinking of air shafts, in 1872 a seam of admirable quality coal had been reached and mining commenced. This was regarded as being very beneficial to the town of Pontefract, a distance of just under 6 miles from Normanton, and in an area of South Yorkshire where coal mining has long since had a chequered history. In the late nineteenth century, coal was a vital commodity for manufacturing businesses and it is probable that John Rhodes was known well by a number of West Yorkshire textile manufacturers. He did appear to have similar interests to Isaac Smith in charitable works and in contributing to public services, but he did not share his Liberal political affiliations.

There are accounts of John Rhodes becoming ever more involved in public services as his financial circumstances improved; for example, he is recorded as being elected as a North Bierley Union Guardian of the Poor in 1863, attending charity balls in Leeds in 1874 and attending meetings of the Pontefract Temperance Mission in 1878. He became involved in local political life from 1879 onwards and was elected as the Conservative member for the West Ward of Pontefract in 1879.

He was to continue to hold local political positions throughout most of his long life, being elected as an alderman for Pontefract and as Mayor of Pontefract on no less than eight occasions, with the eighth time being in 1889. He is recorded as being a justice of the peace for a similar time period and, for a time, held the post of chief magistrate for the borough. Records show that he continued to be very active in his colliery management throughout the period of his part ownership of Salts Mill and Estates and, in 1894, the Prince of Wales Colliery was sinking a new shaft; he was dealing with a serious workers' dispute in 1896 and making a new agreement with the trustees of Pontefract Park for the whole of the 'Silkstone' bed of coal under the park in 1898.

John Maddocks (1840–1924) was not Bradford or Shipley born and bred, and from his birth until his marriage in 1860 he lived in Cheshire. The 1861 census records him as having moved to Chorlton-upon-Medlock in Manchester and working as an assistant salesman in drapery. By 1879, however, he was living at 7 Parkfield Road in Bradford and in the 1881 census was a stuff merchant, living with his wife and children in Manningham with one servant. He had an interest in art and by 1883 was president of the Bradford Art Guild, becoming chairman of the Yorkshire Union of Artists five years later. There are indications that he had established a manufacturing company in Bradford in 1882, which was restructured in 1890 as a limited company named the 'Bradford Manufacturing Company Ltd'. At the time of the restructure it was stated in a report by the *Leeds Mercury* newspaper that the company had a large circle of foreign customers in addition to its home business and the trademark 'Belwarp' had been adopted for its 'morning and evening wear; travelling and tourists' suits and boys hard wear'.

There are few other records readily available about this company; it was not a manufacturer of cloth as the name might imply, but rather a tailoring business and supplier of clothing products. The limited company formed in 1890 was recorded as being set up at the cost of £33,000 and was authorised as having capital assets of £41,000 – significant amounts at the time. As such, the connections between John Maddocks' company with suppliers of cloth will have almost certainly have included Salts Mill (as a major textile manufacturer) and the interest in involving him as a partner in the consortium may have been his success in making and selling the 'end products' for the textile trade, as well as his having 'many foreign customers'. John Maddocks himself also has some involvement in public life, being appointed a Liberal ward councillor for Bradford North in 1888 and appointed as a magistrate for Bradford in 1892. He will have been known to Isaac Smith through the Liberal Party at the very least. He continued to have an interest in collecting and exhibiting artwork and, by 1898, had moved to what was a desirable residential area of Bradford at Park Drive in Heaton – facts that tend to confirm his not inconsiderable commercial success. He continued to chair the Bradford Manufacturing Co. Ltd throughout his time as a partner in the Salts company and beyond this.

These three members of the consortium brought to the partnership their experience in worsted manufacture and in selling the finished products of textile manufacture, as well as some significant financial resources to invest in Salts Mill. The fourth member of the consortium, James Roberts (1848–1935), was well established as a wool and top merchant at the time the consortium was formed and his success in this business made him an ideal candidate for enhancing the supply of the raw materials required by Salts Mill if it was to return to profitability. His early life was not one of privilege and, in fact, his beginnings were very humble. He was born at Oakworth (an area of Keighley) and was the eighth surviving child of James Roberts (senior) and his wife Jane (née Hartley). His father was a weaver by trade who had initially lived and worked at Thornton in Craven, near Skipton, and had probably been a hand weaver. The family had moved to Belle Isle in Haworth by 1861, as this was where James (senior) found work in the mills as a power loom weaver. Neither of James Roberts Junior's parents were literate and with a family of eleven surviving children – six daughters and five sons – they would have had little in the way of spare resources (although James senior did end his working life as a small farmer of 16 acres).

David King's work on the family history of Sir James Robert (Bart) JP, LLD, published in 2012, gives a detailed account of the family history and notes that, despite their own lack of education, James's parents were determined to do their best to give their son as good an education as possible. Although the family were of the Baptist faith, they enrolled James into the Anglican church school at Haworth that had been established by the Reverend Patrick Brontë, father of the famous Brontë sisters. There are accounts that the young James did meet and speak to Charlotte Brontë on a few occasions when passing in the street. There is also

Cottages at Belle Isle, Haworth, today.

evidence, from his own recorded speech when opening the new schools at West Lane Baptist Church, that he had received instruction at that place in his boyhood. King comments that perhaps he attended both schools at different times or perhaps his instruction at West Lane was purely religious. As was common for the times, James commenced part-time work at the age of eleven years, as a worsted spinner at Old Oxenhope Mill, and this is confirmed by the 1861 census. James Roberts was to rise rapidly within the industry and, by the time of his marriage in 1873 to Elizabeth Foster, of Harden near Bingley, he was a mill manager. In 1874, he was recorded as a manager for J. & W. Hodgson & Co., worsted spinners of Legrams Lane, Bradford, and as living with his family at Wickham Street in the Listerhills area of Bradford; he had also become a father to his first son, James William.

King reports that the Roberts's family tradition has it that his wife Elizabeth's father had made it a condition of his agreement to their marriage that James should start up his own business. Shortly after his marriage he had indeed set himself up in partnership with his cousin Joe Feather (a son of his father's sister Ellen) and formed a new Bradford company.

The new company was formed as a 'wool and top merchant company' and went under the name of James Roberts. Beginning at premises in Palmerston Buildings, Manor Row, Bradford, the firm then built a warehouse, Colonial Buildings in Sunbridge Road, Bradford, around 1888 and soon achieved commercial success. James had become aware of vast flocks of merino sheep

Sir James Roberts.
(© Saltaire Archive)

being reared in Russia and had recognised the opportunities to be gained from purchasing merino wool directly from there. He travelled to Russia frequently and to many other parts of the world to ensure that he could obtain wool at the best prices. His travels in Russia caused him to form a strong attachment to that country and he became fluent in the Russian language. As King also notes, 'being a shrewd businessman [Roberts] also began to process his wool on the continent to disguise its true origins, which enabled him to virtually monopolise his Russian sources for a time'.

By 1880 James had had a new house built for himself and his family at Park Road Bingley that he named 'The Knoll' (not to be confused with The 'Knoll' at Baildon, which was the residence of Charles Stead at this time). The 1881 census records James as a wool and top merchant, living at The Knoll, Bingley, with two daughters and two sons (one daughter having died in infancy), a governess and two servants. In a similar manner to Smith, Maddocks and Rhodes, James Roberts continued his wool and top business for some years alongside his shared ownership with these three men of Salts Mill and Salts estate, and it is clear that he brought not only a different area of expertise to this partnership but some significant business acumen.

It is difficult to find accounts of the nature of the relationships between these men but the West Yorkshire Archive Service hold the original directors' minute

The Knoll, Bingley – now Claremont.

books for Salts Mill, dated from 1896 (earlier directors' minutes from the time of purchase of the business in 1893 are not held), and these show that, between 1896 and 1898, Isaac Smith was always present at board meetings as chairman, alongside John Rhodes and James Roberts, with a secretary Mr C. A. Briggs. The fourth board member, John Maddocks, was in attendance far less regularly and was often in Leeds, London, New York and other international cities, either appointing or paying commercial agents. James Roberts was always present but, in the earlier minutes of the period, he was not recorded as proposing items for the board to consider, but was largely noted as a seconder to the proposals of Smith or Rhodes. The brief minutes record many issues relating to property repairs in the village, or letters from 'Mrs Salt' (and later Edward Salt) seeking confirmation of interest to be paid to legatees and annuities. They also carefully list the annual charitable subscriptions to a range of local bodies that were initially supported by Sir Titus Salt and the Salt family (possibly one of the conditions of their original purchase was an agreement to uphold the founder's wishes in this respect). They also cover a number of issues in relation to notices to quit allotments or to plans to sell land owned by the estate (behind Exhibition Building, on the outskirts of Saltaire or in Baildon).

Bertram Foster Roberts.
(© Saltaire Archive)

The preponderance of issues recorded at this time relate to management of the whole estate and, to some degree, the legacy requirements of the estate far more than they do to the manufacture of worsted cloth. But emerging changes are sometimes indicated; for example, the record of an increase in salary for Bertram Roberts from £100 p.a. to £150 p.a. in November 1896 – the first indication that James Roberts's second son had become engaged in the business.

Curiously, while these minutes always record the company bank balance, the amounts owing to creditors and sometimes the amount of reserves, they always refer to a number 1 account and a number 2 account, with no explanation as to why the accounts are split in this way or whether they are required for different business or estate management purposes. Any explanation may have been recorded in minutes covering the period 1893 to 1896 but are probably now lost for all time. The accounts are often at quite low levels – between £3,500 and £20,000 – and, when they are, unusually, recorded as 'audited' in April 1898, the number 1 account holds £7,319 and number 2 account holds £29,288 – higher than the average sums in the preceding two years. Prior to this audit, in 1897 there are a number of points of note recorded in addition to the increase in salary for Bertram Roberts. The first is an agreement to hold a shareholders' meeting to consider the balance sheet; the second is to use an extraordinary general meeting to split 'A' and 'B' shares and adopt new regulations; and the third, dated June 1897, is to seek a guarantee from Mr Maddocks that no loss will be made in his department from 30 September 1897 onwards. There is a sense that a desire was coming from one or more of the partners for greater openness and accountability in the firms' affairs and that perhaps one partner – Maddocks – was failing in his area of responsibility.

In addition to the points above, James Roberts begins to feature in a more prominent role. In February 1897 he was authorised to sell land from the estate in Baildon; in September 1897 he was asked to act as secretary for some board meetings. He also took responsibility for commissioning repairs to the vital river bridge at Saltaire and he had commenced the sale of land behind the Exhibition building. In November of the same year, agreement was made to increase Bertram Roberts's salary to £250 p.a. and, oddly, it is recorded that the 'Third Bradford Equitable Benefit Building Society' sold back 'A' debentures valued at £20,000 to Mr Rhodes. It is difficult to judge what these noted actions and agreements indicate but, early in 1898, the minutes record that Mr Maddocks intended to sell all his shares to the other partners and, on 4 April, a record was made that 'all accounts, agreements and expenses incurred or entered into by Mr Maddocks with Messrs F. Mommer and others are to be transferred to his private account'.

This was swiftly followed by Maddocks selling his shares in equal proportions to Rhodes and Roberts only, but not to Smith, causing Maddocks to hold insufficient shares in the company to retain his seat on the board, which was then declared vacant. Interestingly, the same minutes record that the coatings, dyeing and finishing offices in Bradford and New York were to be discontinued by the Salt company. Had Maddocks been mixing up his separately owned merchanting business interests in the manufacture and merchanting of clothing with the business of Salts? Was he pressured into relinquishing his shares? Whatever the answers may be, at the same May 1898 meeting James Roberts was appointed manager of the manufacturing and merchanting end of the Salt business.

These events were immediately followed in June 1898 by a letter to the board from Isaac Smith protesting strongly about 'the manner in which items of interest, paid to the Bradford Banking Corporation, had been dealt with and the resolution made in his absence in May with regard to share values'.

No comment is recorded as being made by Rhodes or Roberts in response to this protest. In July 1898, however, Mr Bertram Foster Roberts was appointed as a director to fill the vacancy left by Maddocks and, at the same meeting, it was noted that Mr Isaac Smith had transferred the majority of his shares to Rhodes and Roberts (with a small number allocated to Mrs Esther Smith). Another board vacancy was declared and Mr Rhodes was appointed as chairman of the board at a salary of £250 p.a. At this meeting payments of £30,000 each were also made to Rhodes and Roberts, and recorded as 'being entered into the books of the company'; it is also decided that there will no longer be the practice of having a no. 1 and no. 2 account – 'these are to be consolidated'.

Had Smith resigned due to his close family ties with Maddocks and perhaps a view that Maddocks had been treated harshly? Had Smith himself been less than open about company accounts or other matters? These are areas worthy of much more research that could perhaps provide fuller explanations in the future. What is clear is that, during their time as equal partners in Salts Mill and Salts estate, Rhodes and Maddocks had new streets on land behind Exhibition Building named after them that continue to carry their names to the present day (while Smith bought land in the same area, there is no street named after him). Their selling of their shares in the business to the remaining partners in 1898, five years after the purchase of the business and the estate, was an odd affair. Both were to continue with their separately owned businesses after this event, Smith with much greater success than Maddocks, who was explaining to his shareholders why no dividends could be paid in 1912, having sold his private collection of pictures at Christies in 1910 and his London residence in the same year.

From 1898 the owners and managers for Salts Mill and estate were now John Rhodes, James Roberts and his son Bertram Roberts, and, from August of that year, the minutes began to record matters that reflected the business of worsted manufacture to a greater degree. For example, decisions were made to purchase

Rhodes Street, Saltaire, today.

new spinning and wool washing machinery and there was a movement of debts to the new 'profit and loss account'. What is very noteworthy from this point on in the company's affairs is that the bank balances increased significantly – for the most part being between £80,000 to £130,000 – a major difference to the 'consortium years'.

In July 1890, John Rhodes was recorded as voluntarily resigning as chairman of the company and James Roberts was appointed as chairman in his place that August, receiving a bonus of £1,000 in addition to his salary. By September 1890, at a meeting where only James and Bertram were present, extensions and alterations to the wool warehouse were agreed. In December 1890, James Roberts was reappointed as the managing director of Salts at a salary of £2,250 per annum and awarded a tax-free bonus of 10 per cent of the company profits from 1890 onwards. John Rhodes was clearly very unhappy about this and gave notice that he would ask for this decision to be rescinded at a later board meeting. In the event, he decided not to pursue this matter and the mill continued to progress in profitability, recording a bank balance of £130,866 that was confirmed at an audit of the accounts in April that year. John Rhodes attempted one further protest about the bonus payments to James Roberts, but was overruled. In June 1901, Roberts's wife Elizabeth was appointed as a director of the company.

Lady Elizabeth Roberts. (© Saltaire Archive)

It was clearly a key moment for John Rhodes and, by January 1902, the director's minutes record that he had transferred all his shares in the company to James Roberts. The consortium period was completely over and from this point on James Roberts, then aged 54 years, and his son, Bertram, with his wife Elizabeth are sole owners and managers of Salts Mill and the Salts' estate.

In April 1902, the Articles of Association for the company were changed so that the numbers required for the board should be 'not less than two directors' and Mrs Elizabeth Roberts resigned at this point. The scene was now set fair to establish a Roberts' family dynasty with the potential for continued ownership and management of Salt Mill and the Salt estate long into the future.

Had James Roberts bided his time carefully to expose some unfortunate business practices by his early partners in order to gain his family's outright ownership? Had he concentrated on Salts Mill business more effectively than the other partners and devoted himself solely to Salts? The answer to the second point is that he had in fact continued his wool and top merchant company until he became sole owner of the Salt business and estate; it was at this point in 1902 that he dissolved his partnership with Joe Feather (who continued as a top merchant until going bankrupt in 1910).

He did engage with the social life of Saltaire and, as King notes, was active in local political and public affairs. At various times he was a member of the Shipley Urban District Council, eventually chairing this, and was elected to the West Riding County Council and the local rivers board. He was president of the Shipley Textile Association, attended the House of Commons to explain Shipley's opposition to incorporation with Bradford Council to a select committee and became a justice of the peace for West Yorkshire. He was a governor for the Salt Schools and became president of Shipley Golf Club and the Saltaire Cricket Club, presenting the cricket club with a new pavilion in 1914. He was a firm believer in the principles of free trade and was instrumental in the formation of the Bradford and District League against Protection. For a time, he was also one of the joint owners of the *Bradford Observer* newspaper but his public, political and social life was not perhaps as widespread or prolific as that of Titus Salt Junior or Charles Stead. He had declined an invitation to become the Liberal candidate in local parliamentary elections in 1911, the invitation indicating that he had strong political support within the Liberal Party.

As the Salts business grew ever more profitable he seems to have walked in the footsteps of Titus Salt Junior and Charles Stead in respect of his family residences. It may have been no accident that he called his house in Bingley 'The Knoll' and, shortly after the purchase of the business, he took up residence at 'The Knoll' in Baildon – the prior residence of Charles Stead. In 1903 he purchased the Milner Field estate from Catherine Salt, the widow of Titus Junior. In the King's birthday honours of 1909, he was created a baronet and became Sir James Roberts, first baronet of Milner Field, and he celebrated the occasion by giving his workers at the mill a week's holiday with pay. His workers presented him with an illuminated address in a decorated silver gilt casket. Sir James also announced the creation of a pension scheme for those who had worked in the mill and were over sixty-five years of age. To ensure the baronetcy had a stately home, Sir James purchased an estate in Scotland from Lord Perth in 1910. The estate had 7,322 acres and included Strathallan Castle, the Dower House of Machany, salmon and trout fishing in the River Earn, and grouse shooting.

A major question remains: did he succeed in establishing a family dynasty to own and manage Salts Mill and the Salts estate, as well as providing an heir or heirs to the baronetcy and the family 'seat' at Strathallan? The answer is full of tragedy and drama. His first son James William, known as Willy and born in 1874, was to suffer ill health for most of his short life. Willy had been educated at Bingley Grammar School and the Institute Pietsch in Dresden; he studied Russian and German and was noted as being skilled in mathematics. When

The casket, containing a scroll; it was presented to Sir James Roberst by Salts Mill workers.

and how he developed tuberculosis is not clear but he was sent, aged sixteen, to Melbourne, Australia, with Joe Feather (his father's business partner) on a wool-buying trip in the hopes that Australia's climate would help him. This was not to be the case and doctors recommended his return to England. The 1891 census records him as living at 'The Knoll' in Bingley and working as a wool trades apprentice. In 1892 he was sent abroad to South Africa, once again in the hopes that the warmer and drier climate would be beneficial to his health – the *Bradford Observer* reported that it was hoped that he would take up permanent residence there. Unfortunately, he caught a chill and his ill health was such that he was rushed back to England again where, after a bout of pneumonia, he died aged twenty-four years on 3 June 1898 – a point in the history of Salts Mill when his father was gaining ascendancy in controlling the business. He had not married and had no children.

Bertram (Foster) Roberts, born in 1876, was heavily involved in the business by this time and looked set to inherit his father's business acumen and abilities. The younger of his two surviving sisters, Alice Maud Roberts, who was born in 1881, certainly inherited her father's determination, as her life was to be one of adventure and romance; however, it was also such that it was also to later create another family tragedy.

She was educated at a school for young ladies in Switzerland and in 1899 was taken to Paris by her father, where she received a proposal of marriage from a Polish officer with a commission in the German army. Her father approved of the match but Alice was not impressed and refused the offer. She may have already had her sights on a young medical student, Norman Cecil Rutherford, who was studying at Edinburgh and was a year younger than herself. Norman was the son of Doctor John Rutherford, a general practitioner working in Shipley, and it is possible that he was the Roberts's family doctor. He was certainly well known to the family because he, his wife and two sons attended the wedding of the Roberts's oldest daughter, Lily May, in April 1902. Just four months after the wedding, Alice eloped with Norman and got married in August 1892 at the parish church of Ferry Port on Craig, situated on the River Tay opposite the town of Dundee. The parents of neither the bride nor groom were present and the two witnesses to the marriage were unrelated people. In 1903 Norman, having joined the Royal Army Medical Corps, set sail for South Africa and Alice joined him there shortly afterwards. Alice had her first two children in South Africa and returned with Norman in 1907, initially to live in Scotland and later in Ireland. She was to bear four more children to Norman and they had a total of three sons and three daughters; but tragedy and a scandal that caused a national media sensation were awaiting them in 1919, when her husband shot and killed a man he thought was her lover.

Bertram was himself to marry on 22 April 1903 and, as the first of James Roberts's sons to marry, this wedding was a very grand occasion. His bride was Eliza Gertrude Denby, the only daughter of Ellis Denby (later Sir Ellis Denby) of Wycliffe House, Shipley. Eliza was the great-granddaughter of William Denby, founder of the well-known local textile company of William Denby & Sons, with establishments at Tong Park, Baildon and Well Croft Shipley. The marriage took place in the Saltaire Wesleyan Chapel, and Victoria Road was garlanded throughout, with decorative masts and shields erected. Festivities for the villagers and Salts Mill workers were organized in Saltaire Park, with music and tug-of-war competitions between mill departments. It was reported that 4,000 people were served with food. A reception for invited guests was held at Wycliffe House and, among the many expensive gifts for the young couple, was a grandfather clock worth 100 guineas, given by the workforce at the mill. The young couple took up residence at The Knoll in Baildon, recently vacated by Bertram's parents. Two daughters and two sons were born there between 1904 and 1909.

Just fifteen months after Bertram's marriage, however, yet another tragedy struck the Roberts family. In August 1904, during a family holiday in Ireland, James Roberts's youngest son John Edward, born in 1893 at 'The Knoll' Baildon and known as Jack, was fishing off some rocks at Ramore Head, Portrush, near a place known as 'The Washtub' when a witness heard a splash and turned around to see that the young man had been swept into the sea. He was eleven years old and a nearby fisherman named Joseph Fearon leapt into the sea in an effort to save him, but the weight of the boy and the power of the sea defeated him and Jack was drowned. James Roberts and his wife Elizabeth had been staying in Portrush on the north coast of Ireland for a two-week holiday and, as they were leaving the next day, had gone to say their farewells to friends and acquaintances in the area, but Jack had preferred to go fishing with his friends. Jack's parents must have been distraught at the loss of a second son in such an untimely manner; Jack's body was never found. Having lost his oldest and youngest sons through ill health and a tragic accident, James Roberts had some consolation in the fact the Bertram was doing well in the business at Saltaire and that he had another son alive and in good health – Joseph Henry (Harry) Roberts.

The consolation was not to last very long. Bertram had taken on greater responsibilities for managing Salts Mill during these years and his father, who was approaching his sixties, was happy to leave the day-to-day business of running the mill in Bertram's capable hands during his frequent visits to the Strathallan estate from 1910 onwards. Bertram also began to look for a more substantial residence for himself and his growing family and, in 1911, he purchased the 3,200-acre Kilnsey estate in nearby Wharfedale. The estate included Chapel House, a number of cottages, farms and moorland and a licensed premise, The Tenant Arms. Messrs John Harper & Sons, auctioneers, had been responsible for the sale of the estate, which produced an annual rent of £1,537 16s 6d. Bidding started at £25,000 and the figure rose rapidly to £29,000, but it appeared that the bidders were not inclined to go above that sum. The lot was therefore withdrawn but Bertram immediately consulted with the auctioneer, offering £31,000 for the estate, and the auctioneer announced that, 'as the mortgagees were desirous of clearing it as a whole', he would knock down the property to Mr Roberts. Bertram had previously purchased the Stockdale estate near Settle and had become a substantial landowner in the Yorkshire Dales. It is interesting that descendants of Bertram still reside at Kilnsey and manage a visitor attraction and fishing lakes on the site.

Everything seemed set for Bertram and the last remaining brother, Joseph Henry, known as Harry, to continue to build on their father's success. This was not to be the case. Bertram, who was educated at Bingley Grammar School and Harrogate College, had been a keen footballer for teams at Bingley and Saltaire. In 1910 he had begun to suffer from back pain. He was diagnosed as having symptoms of 'serous neuritis' (possibly inflammation of the spinal fluid) and his health began to deteriorate. In November 1911, he was thought to have suffered a 'nervous breakdown' and, after a short rally, he died on 11 January 1912, aged thirty-six. It is highly unlikely that a 'nervous breakdown' could have been the cause of death, but diagnoses of the causes of ill health were far more difficult then than they are with today's armoury of diagnostic techniques and range of sophisticated body scanning methods; it is possible that the symptoms he had were an indication of bone cancer or a similarly life-limiting condition.

Bertram, like his father, had begun to play a prominent role in local public and political affairs, having a seat on the Shipley Education Committee, working as a Salt Schools governor, being elected as a member of the Governing Board of Bradford Royal Infirmary and a president of the Shipley Board of Health – his future had seemed very bright. His funeral took place on 16 January 1912 and his body was laid to rest in Nab Wood Cemetery, Shipley, close to the grave of Willy, his older brother. Sir James Roberts was later (1920) to rename Saltaire Park as Roberts Park in his honour, and there is a small plaque situated on the coach road entrance to the park that commemorates him. This is the only public acknowledgement of the Roberts family's success in rescuing the previously failing mill business, caring for the Salt estate, continuing the charitable payments to a wide range of good causes established by Sir Titus Salt (Bart) and continuing the payments of interest to the Salt legatees. Sir James Roberts now had one remaining son who could participate in the ownership and management of Salts Mill and the Salts estate.

Harry Roberts, born in 1887, had been appointed a director of Salts Mill on 17 September 1908 at the age of twenty-one years, after a period as a departmental manager in the mill. Some of the Salts board director's meetings record that only Bertram and himself were present to make decisions on behalf of the firm while Sir James was seeing to affairs at Strathallan. Harry became joint managing director of the business with his father after the death of Bertram, and the two continued with admirable stoicism to run a successful business until the advent of the First World War in 1914 – a war that was to be instrumental in preventing the continuation of their partnership. At the beginning of the war, Harry was aged twenty-seven years old and liable to be conscripted to fight in France. This would have left Sir James, who was now in his late sixties and in poor health, in a very difficult position – forced to take sole responsibility for the running of the company at a time when his trade with Russia and Germany would be badly affected, while at the same time perhaps being asked to be more productive to help the war effort. The advisory committee and the military representative to the committee, whose task it was to advise the local district tribunals in deciding who was to be 'called up', had initially informed Sir James that Harry would not be liable for service.

The war had been an increasingly bloody and murderous affair and the loss of hundreds of thousands of men led to greater demands that all should fight. When Harry's case came before the Shipley Recruiting Tribunal, he was not regarded as being in a reserved occupation and he was called up. At the first public meeting held by the tribunal in February 1916, at the District Council Offices at Somerset House, a request for exemption from the war was made by Sir James Roberts on behalf of four key operatives from Salts Mill. The four men were recorded as Joseph H. N. Roberts (Harry, managing director), Harry Griffiths (chief clerk), James Midgely (designing room clerk) and Percy Baker (clerk of works). Sir James himself was not present but his company secretary, C. Briggs, was. The verbatim report of the tribunal states that the chair, Councillor Thomas Hill, and five other members were not impressed by the case for exemption made on behalf of Sir Titus Salt (Bart) Sons & Co. Ltd by Briggs, speaking for Sir James. The chair stated that

> Mr. Roberts must be fully acquainted with the Lord Derby Scheme and also with the fact that all available men are necessary. The Tribunal wonders what reason he had for thinking that the regulations under the Derby scheme did not apply to him.

These and other somewhat sardonic questions were put in public and the decision was made to disallow the requests made by Sir James, although eight others were granted postponements, three were exempted and six others were refused. The whole transcript of this hearing, with many barbed comments from tribunal members, suggest that Sir James may have made some enemies locally. Mr Briggs notified the chairman that Sir James Roberts would appeal the decision, and an appeal hearing was held in March by the District Tribunal and a letter from Sir James was read out. The decision was to forward the appeal to the Central Appeal Tribunal, who received a longer, heartfelt letter from Sir James explaining all the efforts made by the firm to support the war effort and affirming that his son had been keen to serve his country, but had been dissuaded by his father in the interests of Britain's national and international trade. It was all in vain and Harry Roberts was called up.

In due course, Harry received his military training, soon attaining the rank of second lieutenant, and was sent with the Royal Dublin Fusiliers to serve on the front

First World War trench warfare.

in France. After just three months of active service, Harry was seriously wounded in his right leg, his calf having been almost blown away, and he received hospital treatment. A report of 20 September, on the sixty-fourth anniversary of the opening of Salts Mill, stated that Harry was making satisfactory progress in a London hospital. He was not required to serve again but multiple factors had affected his father's wish to discontinue with the family business, in addition to the realisation that Harry would be too infirm to take part in running Salts Mill. Not least of these was the Russian revolution in 1917 – but Sir James Roberts's problems in keeping the mill running started earlier, as the war broke out in 1914. Ian Watson's 2016 examination of Sir James's letters at the time demonstrate that the problems were significant. Watson comments on Sir James's long history of establishing trade with Russia and the fact that Sir James had spent three months each year in Russia in the 1880s, travelling around the large estates in the area of Rostov on Don, where he bought merino wool. Emerson's 1913 work explores how Russia was seen as a growing economic power in the years before the First World War.

A letter from Sir James written in 1914 records that he had planned to make another visit to the country to introduce his son-in-law Fred Aykroyd to his contacts there. The letters of the period also show that, in addition to buying camel hair and cashmere from Russia, the Salt business had established a yarn-export business within Russia and Germany through the use of agents based in Bradford, Germany and Russia. The business was experiencing difficulty with these agents and the purpose of the 1914 visit was to try to take out the 'middle men' and form direct trading links. Sir James had also planned to arrange for a German merchant to have a temporary post at Saltaire to gain experience of the trade. It is not clear whether either plan was successful, but the outbreak of war caused Sir James to cancel a trip to Strathallan and remain at Milner Field to 'stand by' his workers. The outbreak of war had caused an immediate crisis for the mill, because 90 per cent of the yarn it exported had been to Germany and Russia – both countries immediately cancelled all their outstanding contracts. The mill was put on short-time working and, in order to compensate his employees, Sir James suspended the rents for the workers' homes in the village. Another immediate problem was the suspension of normal commercial credit operations, and British banks began to call in loans while also increasing interest rates. The government worked to stop the flow of money, closing the stock exchange, and established a moratorium to protect the banks.

Letters from Sir James to David Lloyd George record his efforts to explain the effects of these measures on his business and he wrote to the Board of Trade for advice on paying bills to various German companies. The mill was also affected by the restrictions placed on imports; one of the specific problems for the textile industry at the time was that Germany was a leader in the field of dye products. These problems coming, as they did, soon after the loss of his most experienced son, Bertram, would have caused many a manufacturer to give up. Not Sir James – he began work then to increase the export of yarns from Salts Mill to the United States of America, and began to make efforts to establish a market for his cloth there. He was also aware and hopeful of gaining government contracts for cloth, where there was high demand in order to produce uniforms for all armed services. He reduced the cost of cloth initially to secure a government contract, making no profit, and did eventually get permission to recommence exports to Russia. However, this was not without its problems, due to poor rates of exchange between sterling and roubles. Sir James also began to establish markets in Scandinavia but suffered many delays in obtaining export licenses. His problems in trading were then exacerbated by the refusal of the Recruiting Tribunal to exempt Harry from serving in the army. It is of no surprise, therefore, that letters written by him as early as 1916 clearly indicate that he was considering the sale of the Salts business.

Although Sir James managed to continue the business successfully, despite his many obstacles, he was not the only person to benefit. He had continued to support all the mill workers financially during periods of short-time working and, in February 1917, he called all his employees into the mill yard to report that he was willing to purchase government war loan stock at his own expense and allow each worker who could afford it to purchase some for themselves in return for a small weekly payment. The first phase of the Russian Revolution began in the same month and, by October of that year, the provisional government of Russia had been replaced by a Communist (Bolshevik) government. Sir James had also made a number of investments in Russia and consequently suffered a huge financial loss, leading him to make many visits to Lloyds Bank in London, seeking reparations that were not secured. His many attendances in the bank were noted by the poet T. S. Eliot, who was employed there at the time, and Eliot makes mention of his forlorn figure in his famous poem, *The Waste Land*. A key source of raw materials and an important purchaser of goods was lost to Salts Mill for all time, and Sir James stepped up his negotiations with a syndicate of Bradford 'Wool Men' to sell Salts Mill and the Salts estate. These businessmen were Sir James Hill and his two sons, Arthur and Albert; Ernest H. Gates; and Henry Whitehead, who did, after protracted negotiations, purchase the business and the estate in January 1918.

Despite these immense problems, the Salt business continued to be highly successful and profitable, allowing insight into Sir James Roberts's character and personal strength during a long period of great tragedy and loss. There are sufficient clues to his character in this narrative that give evidence of his astute ability to spot fruitful business opportunities, his great determination to see things through, his personal fortitude and stoicism and his flair for successfully managing a large business; but, very sadly, his legacy is often forgotten by those expounding aspects of the heritage of the industrial community that is Saltaire. He is more than worthy of some last words that try to define his legacy a little more fully.

James Roberts himself recounted, on his attendance at a presidential address (given by Joseph Wright to the Salt Schools in Victoria Hall) on 5 October 1901, that, as a boy, he had walked with his friend from Oxenhope to Saltaire to see the wonderful town with its four lion statues and he and his friend had been met with kindness from a local shopkeeper when they were hungry. Expressing this in public reflected the length of time that he had known about the 'famous' Saltaire and his interest in it. His actions throughout his time at Salts confirmed his huge respect and admiration for Sir Titus Salt; for example, he had not felt the need to change the name of the business to that of his own family, had sustained the annual payments to all the charitable bodies either initiated by or supported by Sir Titus Salt and was instrumental in ensuring continued payments of interest to Salt's legatees over twenty-five years. In 1903 he commissioned a large statue of Sir Titus and had this placed in a prominent position in the then Saltaire Park, where it stands to this day.

He continued the legacy and vision of Sir Titus in treating the mill workers with many kindnesses, and showed great understanding of their needs when times were hard. He had offered to build a new hospital in Saltaire in 1906 (some conditions

Sir Titus Salt's statue, Roberts Park, commissioned by Sir James Roberts.

he set did not find favour with the hospital charity trustees) and he had extended the mill itself by almost a third of the original floor space on the north-west of the original complex, involving the construction of a multi-storey building for spinning that connected with the existing dyeing and finishing plant.

His philanthropy and business improvements did not end there, however, and his charitable actions applied to much broader issues and wider geographical areas. For example, in 1910 he provided a scholarship of £40 for three years to Bradford Grammar School and purchased a large property at Town Well Road in Harrogate as a gift from himself and his wife Elizabeth to Dr Barnardo's, to be used as a home for invalid children, with accommodation for forty-five 'cots'. In 1914 he subscribed generously to the Prince of Wales Relief Fund, provided a pavilion for the Saltaire Cricket Club and offered to place Strathallan Castle at the disposal of His Majesty's Government for the care of wounded soldiers. In June 1916 Sir James gave a gift of £10,000 to Leeds University for the foundation of a professorship of Russian language and literature, and he gave generous donations to the French Red Cross. In 1920 he gifted Saltaire Park to Bradford City Council to hold all its facilities in trust and maintain this for the local community, on condition that it would be renamed 'Roberts Park' as a memorial to his son, Bertram. In 1928 he purchased the Haworth Parsonage and gifted this to the Brontë Society to hold in trust in memory of the remarkable Brontë sisters; this is well maintained and immensely popular with visitors to the present day. In addition, he became involved in the public life of Perthshire and is recorded as carrying out many other charitable acts in that region.

With the exception of the renaming of Roberts' Park and the small plaque honouring Bertram, he showed no signs at all of wanting or needing to have his own image or name remembered in Saltaire, a fact that has led to his legacy to be overlooked more often than it is remembered. It is very sad that neither Shipley Urban District Council at the time (incorporated with Bradford since the 1970s) nor Bradford Council have seen the need to create a memorial to him, given that his effort and skill caused the fulfilment of Sir Titus Salt's vision to be secured for a long time into the future. Sir James himself did not seem concerned by this lack of acknowledgement and this is perhaps testament to modesty and lack of personal ego. His life after the sale of Sir Titus Salt (Bart) Sons & Co. Ltd became much more peaceful. He sold Milner Field to one of the new owners of Salts Mill in 1922; he established a base in London at 30 Hyde Park Gardens; he relocated to a residence named Newland Park at Chalfont St. James in 1922; and gifted the Strathallan estate to his grandson in 1925. Finally, he purchased a lovely property, Fairlight Hall, and estate, near Hastings in 1926 where he remained with his wife Elizabeth, often visited by family members. Elizabeth died there in July 1935 and Sir James died just five months later in December 1935.

His legacy needs to be remembered and celebrated much more today than it has been in the past, if for nothing else that, despite enormous hurdles, significant

The Roberts family at Strathallen on the coming of age of James Denby Roberts, 1925.

The gravestone for Sir James and Lady Elizabeth Roberts.

personal tragedies and the trading difficulties during the First World War, he was able to expand the Salt business considerably after its rescue from bankruptcy. At its sale in 1918, it employed 4,000 staff drawn from the Saltaire, Shipley and Bradford areas (in contrast to the 2,000 jobs available when the consortium purchased it), giving impressive numbers of people work and good conditions of employment for many years into the future. The business and the Salts estate were sold for £2,000,000 and his legacy ought to be one that acknowledges his remarkable achievements.

4

The Syndicate Years 1918–1957

The Syndicate of Bradford businessmen who purchased Salts Mill and the Salt estate from Sir James Roberts in 1918 had had protracted negotiations with him during 1917 in respect to the land values within Salts estate; they were not persuaded to pay his original asking price for a number of pieces of land. They nevertheless eventually settled on agreed prices for estate land and for the mill business – paying just under £2 million. The syndicate was comprised of Sir James Hill, his two sons – Arthur James and Albert Hill, Henry Whitehead and Ernest Henry Gates. Sir James Hill (1849–1936) was born at Harden, near Bingley, and was the son of a wool merchant, William Hill. In 1875 he entered into a business partnership with John Reddihough but, by 1891, had established his own business, initially importing wool and later established a woolcombing and top-making business.

He would have known Sir James Roberts as a fellow businessman and competitor before he became involved in the rescue of Salts Mill and the Salts estate in 1892. He was also a Liberal politician, becoming Lord Mayor of Bradford for 1908/09 and was elected unopposed as a Member of Parliament for Bradford Central Ward at a by-election in 1916, following the death of the Liberal MP Sir George Scott. However, he was to lose this seat in the general election of 1918 to a Conservative (Henry Ratcliffe Butler) and did not stand for Parliament again. He was made a baronet in 1917, some eight years after the baronetcy was awarded to Sir James Roberts. At the time of the purchase of Salts Mill, he was aged sixty-nine years and had a very large family business to run. The *Yorkshire Post* article of 24 January 1918 reporting the sale of the Salt business stated that the firm of Sir James Hill & Sons, Bradford, was 'the largest firm of top makers in the world'.

It is not surprising, therefore, that Sir James Hill did not choose to become a director of the Salt business, and it is likely that he saw his family's involvement as providing a very good career opportunity for his eldest son, Arthur James Hill, who was immediately appointed as a director of Sir Titus Salt (Bart) Sons & Co.

Ltd, initially alongside Henry Whitehead and Ernest Gates. This chapter provides information about these three men and then briefly explores the roles of two other men who were to play significant parts in managing the business after it had ceased to become a private limited company in 1923 – Robert Whyte Guild and Sir Frank Sanderson. Once the firm of Sir Titus Salt (Bart) Sons & Company Ltd had ceased to be the company managing the mill and a new public company had been formed, there were many other individuals, who are too numerous to detail in this chapter, who were involved in some directorial or managerial capacity within Salts Mill. After an account of the three businessmen involved in the 1918 purchase, this chapter will focus on the time period when Robert Whyte Guild and Sir Frank Sanderson were the two players with longevity involved in managing Salts Mill between the 1930s and 1957. It will relate their experiences to some of the economic and international contexts for the business at Salts Mill during this period.

Henry Whitehead (1859–1928), appointed as chairman of the board for the company in 1918, was born at Stalybridge in Lancashire, and the census of April 1861 records him living with his widowed mother Hannah and a younger brother George (b. January 1861) at Warrington Street, Stalybridge, where Hannah worked as a linen draper. His early childhood and education cannot have been privileged, having no father from the age of two years, but his mother worked in selling cloth. The many possibilities for employment in the textile industry probably influenced a determination to work in the trade. Family legend has it that he left home aged fourteen and walked across the Pennines to seek employment in Bradford. At fifteen years old, he obtained a job as a clerk with a Bradford firm of Botany Spinners, Messrs Thomas Ambler & Sons; shortly after this, his mother also made her home in Bradford. Although his duties were in the office, he apparently would spend his 'dinner hours' educating himself about wool sorting, and attended evening classes at the Bradford Mechanics Institute. He saw the importance of the mixture yarn trade, which was then just beginning, and set about the task of making himself master of its intricacies.

Aged nineteen years he moved to work for the Bradford firm of Messrs Firth & Renton and by 1881 he was living with his mother and brother at 28 Middleton Street, Manningham, Bradford, and working as a mill bookkeeper. He married Alice Mawson in 1887 in Rotherham, South Yorkshire, and they had three sons and a daughter between 1889 and 1895. By 1891 Henry was recorded as a manager for a worsted spinner, living in Manningham with his family and one servant – an indication that he was becoming successful in the trade. At the census of 1901, Henry was recorded as a worsted spinner employer, living at 4 Park View Road, Heaton, Bradford, with two servants. He was elected as a member of the Bradford Chamber of Commerce in 1903 and was soon deputed to talk to the 'wool trades' section of the Chamber about the issue of vegetable matter, such as fragments of hemp and straw being regularly found in wool tops and affecting the quality of the finished cloth.

His career rise was rapid and, by 1907, the *Yorkshire Post* was reporting his ownership of mills at Young Street, Bradford, and in Heckmondwike. At this point the family had moved to Hawkswood in Baildon, but, in a similar pattern to Sir James Roberts, his life was beset with tragedy in respect to his children. His three sons predeceased him, significantly affecting his own health as a result. His eldest son, Alfred, contracted tuberculosis and died in 1909 in Melbourne, Australia, aged twenty. His travel to Australia because of this disease indicates that efforts to find a hot, dry climate for sufferers were common at the time, if family resources allowed it, in trying to overcome this disease. His second son, George, died a year later after contracting food poisoning and his youngest son, Harry, died in 1919, aged twenty-three, having survived the horrors of the First World War only to succumb to the 'Spanish Flu', which became a pandemic by 1918 and was contracted most often by those aged twenty to forty.

Sir Henry Whitehead.
(© Saltaire Archives)

Nevertheless, Henry was growing in status and personal influence at this point. He was elected as an honorary auditor for the Bradford Chamber of Commerce and as a member of the Conditioning House Joint Advisory Committee in 1915. He was also involved in trying to resolve a shortage of workers in the trade during the First World War as an employer representative on the committee for worsted trade manufacturers. In September 1917 he was one of three gentlemen elected to the board of control for the woollen and worsted industries. This control board had been established by government in 1916 to control wool supplies during the war, but did not operate smoothly at first. As a result of pressures from the textile manufacturers in relation to the difficulties in getting enough supplies of wool to keep mills running and people employed, an employer's committee was established in 1917. At the first meeting of the employer's committee, Henry was elected as deputy chairman, indicating the respect that others had for him and his knowledge. He was also recorded as giving a number of charitable activities and gifts, was a justice of the peace and had established a new men's club at Heaton Parish Church in memory of his sons (in 1911). In these acts he resembled not only Sir Titus Salt & Sons and Sir James Roberts, but followed the pattern of all the successful textile trade leaders. In 1923 he purchased Stagenhoe, a Grade II listed stately home, in the village of St Paul's Walden, Hertfordshire, indicating his desire to join the landed gentry.

His position in the syndicate purchasing Salts Mill would be equal to the others named as the first directors (A. J. Hill and E. H. Gates), but he carried some seniority in years and experience and so was elected as chairman of the board shortly after the purchase of Salts Mill. Arthur James Hill (1876–1935) was born in Allerton Bradford and was the eldest son of Sir James Hill. He was educated at Silcoates Congregational School, near Wakefield – an earlier pupil there had been Alfred Allott, Sir Titus Salt's friend. Arthur was involved in the wool top making (family) firm founded by his father after leaving school. He married Eleanor Duxbury at Blackburn, Lancashire, in 1898. He is recorded as trading as a wool merchant, living with his family and three servants at Craigmore, Marine Parade in Heysham, Lancashire, in 1901. Between September 1901 and December 1909, Arthur and Eleanor had four daughters and, by 1911, the family were living at the Carin Hydro Hotel in Harrogate, although the reason for this change of location is not known. Arthur began to be involved in the West Riding textile industry affairs and was appointed president of the Association of British Wool Buyers in December 1915 when he was aged thirty-nine years. As he was proposed and seconded for this position, a Mr Willey commented on Arthur's personal qualities, stating that Arthur 'displayed great energy and intelligence in his private business concerns' and expressed the hope that he would bring these attributes to the affairs of the association.

Due to this position he was deeply involved in the question of significant delays in the transport of wool to the Yorkshire area from London, and the *Yorkshire Post* of March 1916 reported how the clearances of wool by government had increased from fourteen to twenty-eight days. After a private meeting held jointly with the Colonial Wool Buyers' Association, a deputation that included Arthur reported to the board of control that 'wool warehouses in Bradford were practically empty and men (sic) were going unemployed for the lack of wool to keep them going'.

It is possible that the difficulties in obtaining wool were exacerbated by some of the impurities being found in the material, which was a cause of vexation for spinning firms such as those owned by Henry Whitehead. It is relatively easy to imagine that both men either corresponded or met on a number of occasions about these matters during the years 1915 and 1916. Mark Keighley (*Wool City*, 2007) notes that Lieutenant-Colonel Francis Vernon Willey, a director of the Bradford textile firm of Francis Willey & Co., was appointed by the government as controller of wool supplies in 1916 and remained in that position for the next four years. The new controller was an experienced 'wool man' but the real trouble in connection with state control was caused by the interference of permanent officials and others who knew nothing about the intricacies of the wool textile industry.

Henry Whitehead and Arthur Hill were to be among the men who helped to 'put these meddlers in their place' and enabled the government control scheme to work smoothly. Keighley (2007) also notes that at the beginning of the war, the proportion of West Riding machinery working on army cloth was, at most, 5 per cent but, by 1917, 63 per cent of combing machines were engaged in work for the forces, together with 43 per cent of all worsted spindles and 52 per cent of looms – producing 250,000 yards of khaki uniform cloth a week and 150,000 yards of great-coat cloth.

Arthur was involved in forming a Bradford branch of the National Motor Volunteers to assist the war effort and became a battalion commandant for the volunteers. His age of thirty-eight at the outset of the war and his involvement in textile manufacture would have been the likely factors in his not being asked to serve at the front. In 1917 he was among a notable gathering of trade associations and Chambers of Commerce at the Trocadero restaurant in London where Sir William Priestley MP provided an occasion with dinner where the issues and needs of the textile industry were discussed. Arthur and Henry, through different trade organisations and membership of the Bradford Chamber of Commerce, had been successful in gaining 'the ear' of men in Parliament during a very difficult time for Britain; and this may have been the seed-corn that led these two men into partnership in purchasing Salts Mill. There are some clues to Arthur Hill's ability to manage employees and a *Yorkshire Post* report of January 1918 comments on the very good relationships that the Association of Wool Buyers had sustained with the trade unions where 'they had not even had to go to arbitration on any point in dispute'. These connections and reputations were to mean a great deal for Salts Mill and Saltaire from 1918 onwards. The mill business had gained

Arthur James Hill.
(© Saltaire Archive)

expertise in (wool) top manufacture and worsted spinning, while the third partner in the syndicate, Ernest Henry Gates, would bring knowledge in the manufacturing of cloth.

Ernest Henry Gates (1874–1925) was not a Yorkshire or Lancashire man and was the third of four children born to William and Elizabeth Gates in Effingham, Surrey. His father had married Elizabeth Gent in 1876 in Lambeth, London. The 1891 census returns record William Gates as a bricklayer and his third son, Ernest, as a 'manufacturer's agent'. Little is known about Ernest's early childhood and education but, yet again, he was a man who did not have influential or wealthy beginnings. He married Eva Siggs, in Wandsworth, London, in July 1900, but by March 1901 the census records him as living at 27 Park Mount, Bradford, with his wife and a servant, and he was stated to be a manufacturer of cloth. He had become associated with the wool trade as the London salesman for a West Bowling textile firm in Bradford, and so his change of residence from London to Bradford was not particularly surprising. However, it is somewhat remarkable that, at the age of twenty-five, he had progressed from being a 'manufacturer's agent' to a manufacturer of cloth in Bradford, Yorkshire. It could be said that it was a striking achievement just to be accepted by Yorkshire 'wool men' as he was a man of 'southern origin'.

His first residence at Park Mount, Bradford, was also in the Heaton area; he was a very near neighbour of Henry Whitehead and may have formed a relationship through their close residence locations, as well as through business interests. His is a story that is worthy of much more research into the company records of the various firms for which he was responsible. It is recorded that Ernest became one of four directors of a new company established in Bradford in 1903. The company was registered as H. B. Priestman Manufacturing Company Ltd with a capital of £30,000 in £1 shares, and the business was to be carried on at Brick Lane Mills 'and elsewhere'. Notably the company was to carry on the business of alpaca, mohair, worsted, cotton, flax, hemp, jute and wool spinners, manufacturers and merchants. There was no initial public issue of shares and the other three directors were H. B. Priestman, J. B. Priestman and J. H. Hinchcliffe.

He established his own company, Ernest Henry Gates & Co., around 1909, later taking over provident mills from Isaac Snowden & Sons. By 1911 Ernest was known to be an important worsted spinning manufacturer, living at Myrtle

Ernest Henry Gates.
(© Saltaire Archive)

Grove, Bingley – just 2 or 3 miles' distance from Saltaire – with his wife, one son and a servant, and was also known to be the occupier of Bocking Woollen Mills at Haworth (near Keighley). There are some clues to his involvement in the West Riding Spinners' Federation through his recorded attendance at the federation's meetings at the Mechanics Institute in Bradford, but indications of other involvements in trade organisations are sparse. He has left only a few traces of his rise in the trade between 1901 and 1918, but the reports of the syndicate's purchase of Salts Mill in 1918 comment on his having places of business in Bradford, Bingley and Keighley, and as being associated mainly with manufacturing. The same *Yorkshire Post* report states that the new proprietors of Salts Mill 'form a powerful combination familiar with the worsted industry, from the raw material to the fully manufactured article'.

Ernest purchased Milner Field (the prior home of Titus Salt Junior and Sir James Roberts) in 1922, moving there in December of that year. His wife was disabled and he was known to be very attentive of her needs but, within ten months of the Gates family moving in to Milner Field, she died. Ernest's own time in the residence was also to be very short lived.

The finances available to the owners of the family firm of Sir James Hill & Sons and the firms owned by Henry Whitehead and Ernest Gates were combined to purchase Salts Mill and the Salt estate, and their investment in this venture would, for a short time, prove to be financially sound. In common with all in the textile industries at the end of the First World War, their first task was to rebuild the export business that had had to be largely abandoned during the First World War. The value of wool exports in 1913 had been £37.7 million but all branches of the industry experienced an extraordinary boom in business during 1919 and 1920. There had been an unprecedented loss of production in 'continental' Europe during the war and its recovery was hampered by the low value of the German mark and the French franc. For a short time, this fuelled an increase in export values to £93.63 million in 1919 and a staggering £134.85 million in 1920. The boom was to be very short lived and multiple problems were soon putting pressure on the textile trade.

The Cabinet papers from the National Archives for the post-war period record that the British economy suffered from a loss of productive resources, capacity and change in the structure of international trade and finance, leading to mass unemployment among demobilised servicemen and widespread workers' strikes. Britain had left the gold standard in 1914 and responded by increasing the supply of money to stimulate the economy; nevertheless production was limited due to a substantial decrease in working hours, agreed after the war, and (by 1919) increased consumer demand led to inflation. Early in 1920 the government was more concerned with returning to the gold standard than with high unemployment and inflation.

In Bradford, industrial unrest was becoming more and more acute and, in an address to the Bradford Textile Society in 1920, Mr T. Buttercase of the Bradford

Dyers' Association stated that there seemed to be a suspicion among workers that 'labour was not getting its share of the profits' – a concern held always by Sir Titus Salt in the 1840s. Mr Buttercase pointed out that that workers had a dread of unemployment, and harboured fears that improved production techniques meant more work for fewer men. In addition, higher prices were being asked for the machinery required for the textile trade – for example, a loom cost £60 before the war but at its end the price had risen to £300 – and this was compounded by delays in getting new machines, which were often being exported to foreign markets. By 1921 most mills were so short of work that their owners were prepared to sell goods at prices below the costs of production. A miners' strike in 1921 added to the problems of West Yorkshire mills, which mostly relied on coal and steam power to function.

Salts Mill did not suffer quite as much in the early 1920s as many of the other Bradford mills, and the directors were involved in a number of deputations to London authorities in respect to the price of wool from Australia and had engaged in a national panel to arbitrate in cases of disputes then prevailing within civil contracts. Arthur Hill was particularly prominent in reporting to an anti-profiteering committee. Henry Whitehead had been knighted in the King's birthday honours list in June 1922 for his national efforts to assist the government both during and after the war; and Ernest Gates had brought his Scottish agent, Robert Whyte Guild, into greater prominence at Salts Mill. Together they had weathered the worst of the regional and national unrest, but the pressures were such that they made a decision to form a public company for the Salt business early in 1923.

On 26 April 1923, the directors announced that 'the famous mills of Sir Titus Salt (Bart) Sons & Co. Ltd are to come under the proprietorship of a public company'. This seminal moment for Salts Mill was to end the ownership of Salts Mill by the company founded by Sir Titus Salt. The directors stated intentions were to 'offer the public a participation in the undertaking' but the decision must have been influenced by the hard times then experienced in the industry and the need for fresh investment. At the time of the first report it was stated that the capital of the company had a value of £4 million but, when the flotation of the 'new' company actually took place in June 1923, the 'nominal' capital value was given as £2,200,000 and the registered name for the public company was to be Salts (Saltaire) Ltd. The June report named the first directors of the public company as Sir Frank Sanderson MP, Sir Henry Whitehead, Mr Arthur J. Hill, Mr Ernest H. Gates and Mr W. S. Robinson. The bankers for the new company were to be Barclays; the brokers, Myers & Co. of Leeds; the solicitors, Slaughter and May and Wade and Tetley, of Wade & Co. of Bradford, and their auditors were named as Price, Waterhouse & Co. of London and the Bradford firm of Paton, Boyce & Welch.

The company from this point became subject to the greater legal requirements of public corporations that included the requirement to publish the names of

The Burling and Mending Department, Salts Mill, Christmas 1936. (© Saltaire Archive)

directors, annual reports and annual accounts – making some records more accessible for the researcher. It is clear from the financial returns for 1923/24 and 1924/25 that the formation of the public company had improved its financial health and the trading profits were £383,115 and £286,636 respectively. Global and national economic circumstances were to worsen, however, and Salts (Saltaire) Ltd was to have its full share of difficult times.

Another director had been appointed for Salts (Saltaire) Ltd by 1924, a Mr Moss Samuel Myers of London (who was already in business in Hull with Sir Frank Sanderson), bringing the board numbers to six people. The limited length of this book only allows for a brief mention of these people and the directors that were appointed between 1924 and 1957; however, Sir Frank Sanderson, a director from 1924 to 1957, and Robert Whyte Guild (appointed initially as a director of Salts in 1929 but having played a significant role in the affairs of Ernest Gates and the Salts business prior to that) were to have roles that require a brief exploration. The initial members of the 1918 Syndicate were to have shorter periods of time in their involvement in Salts Mill. Sir Henry Whitehead continued as chairman of the board until 1927, and in the late 1920s he was involved in a wide range of business and charitable activities – he opened bazaars; attended prestigious trade and commerce dinners; became the chief supporter for the first competitive music festival in Leeds; entertained the prime ministers of New Zealand and Australia during their visit to Salts Mill in 1923; established the Yarra Falls Spinning Mills Proprietary Company (on the outskirts of Melbourne, Australia, alongside

Arthur Hill and Ernest Gates); and took part in the formation of a new private company – the British Model House Company – in 1925 in an effort to promote British fashion. This company opened its headquarters at Regent Street, London, in November 1925, having its first spring collection for show in 1926 in an effort to compete with fashion houses on 'the continent'. An announcement of 17 January 1928, however, reported that Sir Henry had been admitted to a London nursing home to have surgery for an undisclosed condition, and his death was announced in the *Yorkshire Post* on 28 February 1928.

1924 advertisement for Salts Mill goods. (© Saltaire Archive)

Arthur Hill remained as a director of Salts, welcoming the new directors of the board Messrs R. W. Guild, George H. Pepper and Herbert Hey in 1929, after the deaths of Sir Henry Whitehead and Arthur Hollins (who had had a short period as technical director for the board). The same announcement of December 1929 noted that the board had acquired the whole of the share capital of Ernest H. Gates, whose freehold mills comprised of Cross Road Mills, Keighley, Harden Mills, Bingley and Providence Mills, Bradford; the headquarters of both companies were registered as being at Salts Mill. In the 1930s Arthur played a more prominent role in Saltaire affairs, becoming president of the committee organizing the annual Saltaire Conversazione, receiving visitors from New Zealand to Salts Mill in 1930, providing his residence at Denton Hall, Ilkley, for a number of sporting and charitable events, and clearing the debts of the annual Otley Show. Sadly, he became ill in March 1935 and died, aged fifty-nine years, in August of that year.

Arthur Hill had worked hard to sustain the Salts business and had contributed to and influenced trade policies locally and nationally for the benefit of the textile trade. He had been heir to the Sir James Hill baronetcy, but his death preceded his father's by a year. He had outlived Ernest Gates, however, and Ernest's activities had been much more 'political' in arguing the case for textile manufacturers in the hard times of the 1920s. Many newspaper reports of the period give testament to his determination to reduce competition, particularly from France, for worsted and other textile products, which was were seen to be depressing the British market. In speeches to the Bradford Chamber of Commerce, Ernest put forward a motion that all British colonies should impose a duty on the raw products for the trade, with the exception of British supplies. The motion was carried and then forwarded to the Imperial Economic Conference in 1923.

Ernest was a member of deputations to the Board of Trade, requesting that duties be placed on imports of goods from foreign companies and that the government utilise the Safeguarding of Industries Act to protect British products. He strongly opposed free trade at the time (in complete contrast to Sir Titus Salt), writing many letters to local and national newspapers to elicit sympathy for the growing number of unemployed workers, and agitated for a government inquiry into the causes of unemployment, providing detailed analyses of the causes, which were contested by some at the time. Ernest continued to exhibit his goods and those of Salts at national exhibitions, and in 1924 he made an announcement that not only France and Russia would adversely affect the textile trade but that Germany would eventually become a source of stiff completion with Britain for textile products. He showed enormous energy and intellect in his arguments, but failed to see many fruits from these as he was taken ill suddenly in March 1925 when an operation on his foot resulted in erysipelas, causing his untimely death from septicaemia at the age of fifty-one years. He had had remarkable success in his own right within the trade and had become a pugnacious fighter on the company board for Salts.

He had only had one child, a son, Ernest Everard Gates (b. 1903) who, having had an education at Cambridge University, became a Conservative Member of Parliament from 1940 to 1951. Ernest Everard Gates did not become involved in the textile industry or Saltaire.

Ernest Gates was responsible for leaving an important legacy for the Salts business, as he made sure that a man who had worked for him for twenty years, and was highly competent and knowledgeable about the textile industry at all levels, was well positioned in Salts by 1925. Robert Whyte Guild, who had originally been Gates's Scottish agent, was to manage Salts Mill very successfully from the 1930s to the early 1960s, although sadly he is a somewhat shadowy figure in the majority of public records and far too little is known about him.

Sir Frank Sanderson and Robert Whyte Guild were the two men who, after the death of the last founding member of 'the Syndicate', Arthur Hill in 1935, worked in harness and harmony alongside a number of other directors with varying degrees of longevity in the firm. Sir Frank Sanderson (Bart) (1880–1965) was born in Hull and was the third youngest of the thirteen children of John Sanderson and Ann Elizabeth (née Barratt). His father John worked as a colour-manufacture dyer and Frank was educated at Hymers School, Hull. The school was founded following the death of Reverend John Hymers (a mathematician) in 1887, who left a substantial sum in his will for the founding of a boys' school 'for the training of intelligence in whatever social rank of life it may be found among the vast and varied population of the town and port of Hull'. Construction of the buildings was completed in 1893, and many scholarships and bursaries were awarded for the children of families with few means; it has had a history of many prior students becoming successful in their chosen fields.

Frank Sanderson was no exception and he experienced a stellar rise from establishing his own business as a corn merchant when he was aged nineteen years, to then entering a partnership with C. Wray to establish a new limited company in 1907. The company was registered as Wray, Sanderson & Co. Ltd with a capital of £30,000 to act as seed crushers and oil contractors, with a registered office at Morley Street, Hull. By 1911, Frank was living at Analby, near Hull, with his wife, two children and four servants and by 1915 he had become chairman of the Hull Seed Crushers' Association. He also received an important government appointment that year and became Comptroller of Filling Factories – taking 'responsibility for His Majesty's filling factories for bombs and shells' – and was noted on appointment as having a well-regarded business capacity and a great aptitude for organisation.

At the end of the war Frank's business had merged with two other companies and became the Hull Oil Trade Combination, with an authorized capital of £1.5 million. Moss Samuel Myers is listed alongside Frank as a director of this new, large company. Frank moved his family to The Orchard, Maidenhead, Berkshire, in 1919

Sir Frank Sanderson, chairman at Salts from 1924 to 1958. (© Saltaire Archive)

and in 1920 became a director for the Humber Fishing & Fish Manure Company. He was awarded a baronetcy in June 1920 for his work in the First World War and became the Baronet of Malling Deanery, Sussex. In 1922 he commenced his political career at the November general election and gained the seat of Darwen, Lancashire, for the Conservatives. He was to lose his seat in the election of 1923 to a Liberal, F. C. Hindle; he regained it in 1924, only to lose this seat again in 1929. He was eventually provided with the 'safe' Conservative seat of Ealing (later Ealing East) and remained as a Member of Parliament until 1950, where Hansard records show his many contributions to debates, in particular those concerning trade and industry. His appointment as a director of Salts (Saltaire) Ltd in 1923 was probably due to his success in business and politics, but quite how he was known to the first directors is uncertain.

Robert Whyte Guild (1876–1966) was born in Edinburgh, the eighth of eleven children for James and Mary Guild (née Scott). His father was recorded as a linen draper and by 1891 Robert was working as a commission agent's apprentice, which most probably involved the sale of textile products. He married Jessie Wayne in

1901 at the Windsor Hotel, Glasgow, in a religious ceremony conducted by the Reverend David Forsyth. The West Yorkshire Archives hold records that show Robert was working as an agent for Ernest Gates from around 1904 for sales in Scotland and some northern areas of England. It is interesting to note that his 1911 contract with Gates gives his business address as 4 North Court, Royal Exchange, Glasgow, and his commission is higher than that of all other agents employed by Gates for work in Manchester and Lancashire at the time. It is not clear when Robert commenced work for Ernest Gates in a senior capacity in Gates's companies in Bradford but he must have impressed him with his abilities and knowledge. In 1924 he is the only other person recorded by name on an advertisement for Ernest Gates's exhibition of textile products at Wembley, London. There are also private letters between Robert and Arthur Hill in the mid-to-late 1920s, mostly in respect of claims being made for interest payments from a solicitor acting on behalf of the descendants of Sir Henry Whitehead, and Robert also often wrote to Arthur Hill about the outcomes of interviews for middle managers in Salts Mill. The clues in these documents indicate that he was trusted by all three directors of Salts long before he was appointed as a director for the company.

Arthur Hill, Sir Frank Sanderson, William Robinson and Moss Samuel Myers (and for a short time Ernest Gates) had an unenviable task in keeping the Salts business afloat and Salts Mill working throughout the 1920s; in 1925/26 they recorded a trading loss of £46,093 that would have reflected the industrial unrest of the time. British miners were being asked to accept reduced wages and work longer hours, prompting them to go on strike in May 1926. To support them, the Trades Union Congress (TUC) called a general strike that brought the country to a standstill for around five days. The TUC quickly called off the general strike but miners struggled on until November 1926 and had to go back to work for longer hours and less pay. The dates of these disputes do not account for the trading loss for Salts as the losses are reported in March 1926 and relate largely to 1925. It is more likely that the cheaper imports of cloth from Europe and the general public's – with so many unemployed – greater inability to purchase new clothes had had significant impact. The worst was not over in terms of global and economic depression. While Salts' trading profits had returned in 1926/27 and 1927/28, these were to descend to a deficit of £287,978 in 1929/30 – the magnitude of which would have placed many firms into liquidation. In 1929 the United States of America suffered a stock market crash; world trade slumped, prices fell, credit dried up and new taxes were placed on imports of foreign products. The worldwide economic crisis was bound to affect the textile industry and Salts (Saltaire) Ltd.

The directors were taking action and a circular issued to shareholders in December 1929 was to bring important changes to the board and to the company itself. Sir Frank Sanderson was named as the new chairman of the company; Messrs Paton and Robinson retired from the board and new appointments for directorships were announced for R. W. Guild (noted by then as managing director for Ernest H. Gates & Co., spinners and manufacturers), George Herbert Pepper and Herbert Hey. George Pepper is noted as being chairman and managing director of Pepper, Lee & Co. manufacturers and Herbert Hey, of Hey & Co., spinners of Keighley. This move not only brought fresh expertise to the Salts Board but started a pattern of acquisition of other companies by Salts. The same circular announces 'a fusion' of interests by the purchase of the whole of the share capital for Ernest H. Gates & Co. by Salts. No purchase price is recorded in the circular but Gates's three mills will probably not have been valued at a low price, given the context of depression. It is quite possible that it is this purchase (that was to reap many benefits later) that was the main cause of the trading deficit for Salts in 1929/30 rather than global depression.

While trading profits returned to the business from 1931 onwards, the Salts business was experiencing similar issues to those of all engaged in textile manufacture. There was a need to replace old machinery in order to compete in Europe; there was a need to move to electricity as the source of energy for mill machinery and remove the dependence on coal supplies; and, in the views of senior people in the trade, a need to safeguard British textile products. In February 1929, Robert Whyte Guild was a significant participant in the government's Wool Safeguarding Enquiry. He attended sessions held in camera and in public and provided an impressive amount of statistical evidence about the number of looms and operatives that were standing idle at the time, the abilities of British manufacturers to make exactly similar cloth ranges to those imported, and the need to secure bulk business in the home market rather than this going to foreign competitors. He gave a lot of technical information about the wide range of British cloth, provided many samples from Salts Mill and passionately spoke about the effect of duty placed on exports. He gave much more evidence of the factors causing poor trading, and his examination by members of the inquiry lasted around three full days. In his concluding remarks he said that 'everybody in Bradford is most anxious to pull his weight if we can only reach a basis on which we can compete'.

In fact, the 1921 Safeguarding of Industries Act, which had placed duties of 33 per cent on a wide range of goods, was renewed for a further five years in 1926 and extended in scope for those years. It was repealed in 1930 and replaced by the Abnormal Importations Act of 1931. The principal British goods to be protected included linen and woollen goods, and a maximum duty that was initially applied was 50 per cent. The efforts of men such as Sir Frank Sanderson and Robert Whyte Guild undoubtedly played a part in enabling the textile trade to compete more effectively at home. The years between 1930 and 1938/39 gave Salts continued trading profits and were also to mark a number of additional acquisitions of other local companies. In May 1930 Robert Whyte Guild became managing director for Salts (Saltaire) Ltd and the firm of John Wright – with premises at Harden, Cross Roads and Damens (Bingley and Keighley) – was acquired.

Concerns remained within the board of directors about the need to bring in electricity to reduce their dependence on coal, replace old machinery and secure bulk orders for their goods. Managing the part of the Salt estate that comprised the residents' houses, shops in Saltaire village and additional streets of houses on the periphery of the village and having to maintain these properties and parcels of land had become a large burden. This was exacerbated by the high number of rental debts – probably due to the periods of time when workers had been unemployed.

The board made a decision in 1933 to sell the village houses, shops and some of the plots of land and other houses on the village outskirts. The costs of purchase are not known but the purchase was quickly agreed by the Bradford Property Trust, owned by a Mr. Fred Greswell and his partner, Norman Foster.

When Fred Gresswell was interviewed by the *Yorkshire Evening Post* about his purchase, he said that he would not disclose the amount but the deal comprised more than a thousand houses and shops, and he added that the sale involved one of the largest purchases of industrial dwellings that had taken place in the country. This sale effectively ended the vision of Sir Titus Salt of an industrial community. Copies of some of the early conveyancing documents post 1933 to private owners are held in the Saltaire Archive and are worthy of research – over time – in their own right. The directors of Salts (Saltaire) Ltd were aided by the sale of the village, both by removing their responsibility for the costs of the maintenance of the houses and the costs to the company of rent collection. The directors were then able to use the capital generated by the sale to purchase new cap spinning machinery in the mill.

In 1936 Pepper, Lee & Co. company – worsted manufacturers of Wyke, Bradford – was also 'fused' with Salts and, in the same year, Salts bought an interest in an Irish worsted mill in Tullamore, County Offaly, which was renamed Salts (Ireland) Ltd in 1937. Salts (Saltaire) Ltd now owned and managed around seven large worsted processing and manufacturing sites and, while it had ceased to manage the village houses, many village residents were still employed at the mill. Robert Whyte Guild had become affectionately known by mill workers as R. W. Guild and he did become involved in village affairs. Both himself and Sir Frank Sanderson attended the annual conversazione, horticultural and other local social events, and were to receive royal visitors and important commonwealth prime ministers to the mill in the 1930s. R. W. Guild's eldest son, Robert Park Guild, was appointed as president of the committee organizing the annual Saltaire Conversazione in 1933. In October 1937, King George VI and Queen Elizabeth were shown around Salts Mill. In the mid-1930s, R. W. Guild had the foresight to approach Montague Burton who at this point were turning out 35,000 reasonably priced suits each week, requiring 6.5 million yards of cloth each year. Through Guild's efforts, Salts was to become the most important supplier of cloth for Burtons for the next forty years. It was clear to Salts staff that R. W. had all the requisites to bring prosperity back to Salts (Saltaire) Ltd.

George VI and Queen Elizabeth visit Salts Mill in 1937. (© Saltaire Archive)

Given his considerable business interests in Hull and his long political career, Sir Frank Sanderson must have had to spend much time away from Salts. That, coupled with the very few reports of involvement in public affairs for R. W. Guild, give the sense that it was Guild who became the workers' perceived head of the business, the trusted manager who knew how to keep the mill busy. Some past workers in the mill have recorded their affection for him in oral histories of their times employed at Salts, relating tales of his 'knowing everyone who worked in Salts Mill personally'.

The board of Salts (Saltaire) Ltd were soon to face new problems, as the rise of the Nazi party in Germany and Hitler's confrontational leadership within Europe began to make the future ever more uncertain. Salts (Saltaire) Ltd had continued to perform well after 1930 and their average trading profits between 1931 and 1938/39 were an impressive £209,644. The mill had commenced operating two-shift working in some sections and had full employment across its business. The shadows of a second major war were looming, however, and in 1938 every textile mill in Bradford took air-raid precautions, beginning to recruit and train volunteers for future air-raid events. In April and May 1939, several months before war was declared, large government orders were placed for 'fabrics of stout constitution and serviceability'. At the beginning of June, the government issued contracts for 3 million lbs (weight) of wool tops for the manufacture of khaki cloth and, after a stressful period of anticipation, war was declared with Germany on 1 September 1939 by Britain and France.

A system of wool control was set up within days of the declaration of war. The headquarters for this were settled at the Hydro, Ben Rhydding, in nearby Ilkley and Sir Henry Shackleton (a Shipley worsted manufacturer) was appointed as controller. The principal duty for the controller was to regulate the supply of raw materials to spinners and manufacturers, and the decision to place this near to the major textile manufacturers was probably founded on lessons learnt about the effect of delaying supplies in the First World War. For the most part the control systems worked efficiently and the mills were kept supplied, with many replacing male workers with women for the duration. By 1941 a National Wool Export Corporation was formed in Bradford to promote British wool exports, financed by the Board of Trade. Its first chairman was one of Salts' directors, Herbert Hey. The government purchased the entire wool clips of Britain, Australia and New Zealand for the duration of the war. As the war progressed, the government also introduced a 'concentration of production' scheme that was to result in some sections of the textile industry losing around one third of their production capacity.

Salts (Saltaire) Ltd continued to show healthy trading profits between 1939/40 and 1945/46, with an average trading profit of £290,439. The constant pressure on production facilities throughout the war had brought home to many in the textile industry that their mills were out of date and old fashioned, but Salts had coped as well as, if not better than, most under the guiding hand of R. W. Guild. As the conflict began to show signs of defeat for Germany, Salts turned its direction to the post-war need for the return of many workers and the recruitment of those displaced by the conflict in Eastern Europe. In the event, many of the 860,000 textile workers who came back did not return to the industry and this, coupled with the extremely harsh winter of 1947, which affected the transportation of coal and caused much loss to farmers of flocks of sheep, was to bring the industry almost to its knees. Salts had prepared well and recruited significant numbers of Polish, Ukrainian and other European workers and worked hard to make factory conditions more pleasant in order to retain female staff.

In May 1946, Salts also made a decision to open a new factory at Uddington, Lanarkshire, and building commenced for a new spinning and weaving concern of 150,000 feet in dimension with provision for future extension. This new factory was to provide employment for 1,000 staff in an area where unemployment was the problem, while West Yorkshire textile firms had a shortage of operatives. The scaling down and removal of war controls and the return to free enterprise was a complicated process, and the first priority was to re-convert the machinery for civilian production. Sir Frank Sanderson stated at the July 1946 annual meeting that the initiative at Uddington was primarily due to 'the driving force of our managing director Mr R. W. Guild'. In regional terms, the wool stock held by the British government amounted to 449.66 million pounds of wool and many thought it would take a decade to get this supply back into the market; however, it was all sold by 1951 for the sum of £431.5 million. Meanwhile, Salts had acquired the company of John Mitton & Sons, worsted spinners at Clarence Mills, Cleckheaton, at the end of 1950, and was fully back in business by 1953 and preparing for a grand celebration.

1953 marked 100 years since the opening of Salts Mill by Sir Titus Salt: peace had returned to Britain, unemployment was at its lowest for many decades and the Salts business had grown to a sizeable group of companies. The board of directors agreed with R. W. Guild that a great celebration was in order, and that all the workers at Salts Mill should be invited to a day's outing to the popular seaside town of Blackpool. A special commemorative service, attended by the Lord Mayor of Bradford, was held in the Saltaire United Reformed Church, while the mill was decorated with flags and bunting and special trains were commissioned.

1953 also marked the acquisition of W. M. Rennie & Co., worsted spinners of Leeds, and Robert Park Guild joined the board of that company. In January 1954, Salts announced that it had purchased the whole of the ordinary shares of Josiah France Ltd Worsted Makers from within their own resources, and this marked a first for gaining a company that had always been famous for the 'Huddersfield Quality' of its cloth. It had been a long-standing claim that Huddersfield mills produced the highest-quality cloth and this had been borne out several times by the greater success of these mills over the years in exporting products to Europe and beyond. In September 1955 at the 32nd Annual General Meeting of Salts (Saltaire) Ltd, Sir Frank Sanderson announced the appointment of Robert Park Guild as joint managing director for Salts with his father, Robert Whyte. R. W. Guild was seventy-nine years old at this point and was most probably aiming to ensure that his succession would be by his son.

In 1956 and 1957, after great business success and the growth of the Salts group of companies, Salts shareholders were delighted when they received dividends from their shares of 12.5 per cent and 15 per cent respectively. Salts (Saltaire) recorded their biggest turnover of any year in their history – larger than their trading profit of £1,208,261 of 1953/54 could have predicted. The achievements

Salts Mill, decorated for the centenary celebrations. (© Bradford Industrial Museum)

The special train for the centenary workers' trip to Blackpool. (© Bradford Industrial Museum)

Robert Park Guild, standing next to the Lord Mayor. (© Bradford Industrial Museum)

had not gone unnoticed and, in October 1958, the London merchant bankers Singer & Friedlander, acting on behalf of the Bradford worsted spinners Illingworth, Morris & Co., made a surprise bid for four million of Salts' ordinary shares. Illingworth, Morris already owned approximately 29 per cent of Salts' ordinary capital and the offer, which succeeded, gave them around a 60 per cent holding in the Saltaire firm. The transaction marked the end of another era at Salts Mill.

5

The Illingworth Morris Years 1957–1986, and the End of Textile Production at Salts Mill

The Illingworth, Morris surprise transaction to take over the Salts company late in 1957 had marked the end of an era that had begun with Sir Titus Salt's foundation of Salts Mill. The mill, after the Salt family ownership, had survived many changes of fortune but had continued in to be in the 'ownership' of predominantly Bradford-based businessmen and industrialists. Now, while many in the locality knew of this new company's name and its existing textile concerns operating in and around the Bradford district, there was much less knowledge held about its London-based owners. Local knowledge was limited as to when and how Illingworth, Morris & Co. Ltd, was established, and how this company was in a position to take over a very successful business at the height of production levels for worsted manufacture – perhaps especially in the case of the mill operatives employed at Salts Mill.

This chapter will begin by concentrating on the family that had established Illingworth, Morris in 1921 and explore the extraordinary nature of their entry into the textile industry and how they came to grow this business after initially having very different business interests. The chapter will then go on to focus particularly on one man's story of what it was like to work for Illingworth, Morris across a range of their textile companies before the acquisition of Salts (Saltaire) Ltd – a man who, after that date in 1957, was to return to a senior position at Salts. This man is Donald Hanson, who spent much time with one of this book's authors in 2012/13, giving a very personal insight into his experiences of the company that enabled his career to reach its height at Salts in the late 1970s and early 1980s.

The company Illingworth, Morris had been established in 1920 by the London-based financier, Isidore Ostrer. The *Stock Exchange Year Book* states that Illingworth, Morris & Co. Ltd was originally registered on 18 Feb 1920 as Amalgamated Textiles Ltd. At this point, Isidore Ostrer was listed as a director in its annual return but, by the next annual return in November 1921, not only had the company name changed to Illingworth, Morris & Co. Ltd, but Isidore was also then recorded as being chair of the group. The company was listed as a 'holding company' owning all the shares in the following West Yorkshire-based companies:

Bowling Green Mills (Bingley) Ltd; Daniel Illingworth & Sons Ltd; G. H. Leather Ltd; Globe Worsted Co. Ltd (Huddersfield); James Tankard Ltd; Greenwood Dyeing Co. Ltd; Illingworth, Morris, Trading Co. Ltd; John H. Beaver Ltd; John Robertshaw & Co. Ltd; and William Morris & Sons Ltd.

It is probable that Amalgamated Textiles had been the company owning seven of the above companies, with the Ostrers holding the majority of the shares in two of them – William Morris (Sowerby Bridge) and Daniel Illingworth & Sons Ltd (Bradford) – before Isidore was appointed as a director of Amalgamated Textiles.

How he managed, within one calendar year, to become chairman of a group of nine textile companies – changing the name of the group to Illingworth, Morris & Co. Ltd. – is not known. The very difficult economic circumstances of the early 1920s may have made it easy for these rapid acquisitions as so many West Yorkshire mills were struggling to survive at the time. What is clear is that some important West Yorkshire textile companies were then in the hands of 'financiers and investors' rather than 'wool men'. The company began a pattern of acquisitions over many years from that point, both in Britain and within countries such as Germany, South Africa and India. The first additional company to be acquired, in 1923, was that of James Tankard, carrying on business at Upper Croft Mills and Albion Mills, Bradford. These were mills that specialised in spinning coloured Botany yarns, employing around 1,000 workers.

The story of the beginning of Illingworth, Morris, recounted by Isidore himself in later years, was that his Lothbury Investment Corporation (established in 1919) had been responsible for a flotation of shares for two textile companies in 1919. These were Daniel Illingworth & Sons (worsted spinners), Whetley Mills of Bradford, and William Morris & Sons (worsted spinners) of Sowerby Bridge. The flotation failed and Isidore decided to retain ownership of the shares; in doing so, he made his entry into textile manufacture. His story does not include detail of what must have been a 'boardroom coup' within Amalgamated Textiles, but it is indicative of the spending power at his disposal immediately after the First World War.

In 1921 Isidore and two of his brothers, Mark and Maurice, also founded the Messrs Ostrer Bros. Merchant Bank, based at 25–31 Moorgate, London EC2. They commenced a pattern of avoiding direct involvement in the day-to-day management of the textile companies they owned or indeed changing the established names of the individual companies, which would have had a long and probably trusted reputation within the textile industrial community. All these early companies were situated in the West Riding of Yorkshire and were always to be managed 'at a distance'. It seems that Isidore, in particular, satisfied himself with either retaining or attracting suitable Illingworth, Morris board chairs, directors and managing directors who had sufficient knowledge and skill to continue to manage them well. The Ostrer involvement appeared to be restricted mainly to monitoring the balance sheets and keeping a weather eye on profitability, with instructions from time to time about senior personnel. This should not give the reader the impression that the Ostrer brothers weren't capable of spotting rising stars in these companies over time.

A fascinating connection between Donald Hanson, who was to catch Isidore's eye in the early 1950s, and the company of Sir Titus Salt (Bart) Sons & Co. Ltd had begun to emerge as early as the 1860s. In 1861, William Hanson (Donald's great-grandfather) was living with his wife Martha and three children at 40 Whitlam Street, Saltaire, and William is recorded as working as a fitter at Salts Mill. Their children were Rose (b. 1852), Frederick (b. 1854) and Eva (b. 1862). By 1871 the family had moved to what was then numbered as 29 Albert Road, Saltaire – the group of houses in Saltaire village generally used by senior staff and managers working for Sir Titus – and William was recorded as a machine fitter, with his son Frederick as a joiner/carpenter. The Hanson family knew that Sir Titus Salt valued William Hanson's engineering skills highly and that he was personally presented with likenesses of the great man. His son Frederick (father of three sons and two daughters) also worked at Salts Mill, and became responsible for a team of workers charged with maintaining the Saltaire houses. Family legend has it that, one hot August day, when Frederick's team had worked especially hard, Frederick took them to a pub called the Ring of Bells in Shipley for a drink. The outcome of this was that his father William sacked him, and Frederick moved with his family to Slaithwaite, near Huddersfield – an event that must have happened at some time between 1896 and 1897.

What is also of interest is that Frederick had met his wife, Isabella Anderson (who had been born in Orkney, Scotland), through her appointment as governess to Titus Junior and Catherine Salt's children. The Saltaire Archive holds a children's story book given by Catherine to Isabella for 'William and Harry' – two of Frederick's five children.

A further connection between the Hanson and Salt families was to occur through William Hanson Senior's younger daughter Eva, who became a school teacher in Saltaire. She met and got engaged to a book-keeper called Percy Johnson, who had been working at Salts Mill sometime before 1885. Percy must have been a senior employee at Salts because he was persuaded to travel to Dayton, Tennessee, with Titus Salt Junior and Charles Stead in around 1884 to take responsibility for the finances of the Dayton Coal & Iron Company Ltd that they had founded in 1883. Eva travelled to Dayton to marry Percy Johnson in Rhea County in 1885 and, apart from visits back to see family in Yorkshire, they remained in the United States and eventually were to settle in Canada. The harsh action of William Hanson Senior that led to his son Frederick and family relocating to Slaithwaite by 1897 was to lead to at least one of his three sons, Harry, starting work at Globe Mills in Huddersfield. Harry had some success, becoming a manager at Globe, possibly before this mill became part of the Illingworth, Morris group of companies. These early events were the source that was to provide for Donald's much later connections with the Ostrer family and the return of an important member of the Hanson family to Salts Mill in the late 1960s.

Well before Donald caught the eye of the Ostrers, in addition to acquiring a textile group of companies in the early 1920s Isidore was approached by a stockbroker, whose client wanted to buy control of a film business. This proposition had been turned down by various other issuing houses because investors of the time considered film-making to be a risky business proposition. The client was a Colonel Bromhead, whose French partner was Leon Gaumont – the company was the Gaumont Company of London. Isidore decided to buy out Leon Gaumont's control of the English company for £250,000 and formed the Gaumont Trust Corporation Ltd to hold the controlling shares. He initially regarded this purchase as speculative investment and for five years he left Bromhead and his brother to run the company in much the same way as they had done so previously, mirroring his early approach to textile production in the north of England. Isidore Ostrer's ownership of an investment company was quickly followed by the establishment of a merchant bank with two of his brothers, which had the financial ability to retain the whole share issue of a number of sizeable textile companies and also to find sizeable sums of money to buy control of a film production company. This could lead anyone to assume that the Ostrer family had personal wealth of some magnitude, but this was far from the truth at the turn of the twentieth century.

Jacob Ofstrosky (1862–1932), sometimes recorded as Ostravich in census returns, was to become the father of seven children. Jacob had fled Ukraine to escape anti-Semitic persecution in the 1870s, and travelled to England via Paris, where he met his future wife Francesca (Fanny, d. 1932). After settling in London and experiencing the death of their two-year-old first son Nathan in 1885, Jacob changed his name to Nathan Ostrer and the family settled at 16 Great Alie Street, Whitechapel – an area that had become home to many European Jewish migrants. Nathan found work in London as a jeweller's salesman and the family increased with the births of a further six children (Sadie, 1885; David, 1886; Isidore, 1889; Harry, 1890; Mark, 1892; and Maurice, 1896). Little is known about the upbringing or education of the Ostrer children, although it was soon clear that it was Isidore, the third surviving son, who was to emerge as the family member with a strong business and financial acumen. He began work as a stockbroker's clerk in the City of London prior to the onset of the First World War. Isidore was called up to serve in the war but was demobbed before going to the front for health reasons. His older sister Sadie had married by this time and had moved to New York to live with her husband, taking her youngest sibling, Maurice (aged eleven), to live with her – suggesting that the family as a whole were reliant on the fortunes of their children to help with their income and welfare needs wherever possible.

Isidore, meanwhile, clearly began to prosper while working as a stockbroker's clerk in 'the City', but his acquisition of substantial wealth by 1918 is shrouded in mystery.

One family story told is that he was working in the city stockbroking firm on a day that no senior partners were available when their most important client, Rothschild, paid a visit. The family suggest that Rothschild was so impressed by the young man's persona and abilities that he would only be served by him on future visits. However, this in itself would not account for the wealth he had acquired by 1918. Rumours are known to have circulated at times that he had managed to purchase a stock of army uniforms at a low price in the first year of the First World War and make a handsome profit when selling these to the War Office. A more likely explanation, however, is that what is now understood as 'insider trading' on the stock market was not illegal at the time, and Isidore may well have profited handsomely from this practice during his time as a stockbroker's clerk.

What is certain is that he had prospered sufficiently to pay for his sister Sadie and his youngest brother Maurice's return from New York to England with a first-class ticket in 1918 and had set up an investment company followed by a merchant bank.

Isidore Ostrer as a young man.

This was swiftly followed by the acquisition of companies that could not have been more diverse than those engaged in producing textiles and cinematic films. Isidore had married Helen Dorothy, daughter of Lloyd Spear-Morgan, a solicitor and the granddaughter of the high sheriff of Carmarthenshire on 31 March 1914. There is nothing to suggest, however, that Helen had personal wealth. She had been a teacher of ballroom dancing in Wales and there is no evidence of her inheriting land or other assets to support her new husband.

The history of the Gaumont-British Picture Corporation deserves further research; here, some aspects of that history are worthy of note for the way that they provide insight to the personalities and attributes of the three Ostrer brothers, who became prominent in their careers. Isidore was, it appeared, able to pre-empt the market and go on to judge the right timing for expanding his business interests. The prospects for British film-making were opened up by the first Cinematograph Films Act (1927). This Act was designed to stimulate the declining British film industry and it introduced a requirement for British cinemas to show a quota of British films for a duration of ten years. This was accompanied by the introduction of 'talking movies' and it encouraged Isidore to take direct

Mark Ostrer.

control of Gaumont-British, becoming chairman in August 1929, with Mark as vice-chairman and Maurice as joint managing director.

While in these controlling positions the Ostrer brothers created a circuit of 350 cinemas in England, and the brothers turned their hands to film-making. The old Gaumont Studio at Lime Grove, Shepherds Bush, was rebuilt in 1932 (gaining six large production stages) and Isidore orchestrated a merger between the Gaumont and Gainsborough film production companies during 1927, acquiring the Gainsborough Studio in Islington in 1928. Isidore encouraged his production chief, Michael Balcon, to make big budget films with an appeal for an international audience, and Gaumont-British (with Gainsborough Pictures Ltd as a production subsidiary) produced 721 films between 1929 and 1941, engaging directors such as Alfred Hitchcock and employing many British actors who became well-known film stars, such as Dirk Bogarde and Richard Attenborough. In 1932 the company also made a successful takeover bid for Baird Television.

The American market, however, proved to be resistant to British films and ultimately substantial losses were made. Isidore had encouraged American investment in Gaumont-British as early as 1929, when William Fox bought a substantial (although not, as he thought, a controlling interest). In 1936 another deal with the heads of Twentieth Century Fox and MGM was planned, but it was disrupted by a hostile counter bid by John Maxwell of the rival Associated British Cinema. As a result, by 1937, Gaumont-British's production activities were severely curtailed. In October 1941 Isidore cut his losses and sold his shares in Gaumont-British to J. Arthur Rank, although it was not until 1944 that the conflicting interests in the corporation were sufficiently resolved for Rank to assume outright control.

While Isidore played a lead role in acquiring assets, identifying external investors and engaging the senior personnel for Gaumont-British, it was Maurice who chiefly cast himself in the role of film producer. Family sources suggest that younger brother Maurice had wanted the position of production chief held by Malcolm Balcon and, after Isidore sold his shares in the company in 1941, Maurice did have some golden years directly producing films such as *The Wicked Lady*, starring Margaret Lockwood, for Gainsborough Studios. This continued until 1946, when he resigned due to J. Arthur Rank's refusal (on moral grounds) to produce a sequel to the box office success of *The Wicked Lady*. Maurice then moved to Hollywood for a year, hoping to make a career there as a film producer, and, gaining no backers, spent one further year in London as an independent film producer but failed to make progress in this direction.

Ultimately, Mark Ostrer (1892–1958), who had served very effectively as the public face of the Ostrers for Gaumont-British, was the one to remain with the company acquired by J. Arthur Rank and he retained his place on the board there until his death in in 1958, aged sixty-six years. Throughout the years from 1921 to 1941 the three Ostrer brothers had a very active involvement in film production and distribution, and their other male siblings, David and Harry, were also given jobs within Gaumont-British – although not at a senior level. Neither the oldest Ostrer son David, nor the younger Harry, born between Isidore and Mark, held shares in the film company or had any involvement in the textile enterprises within the Illingworth, Morris Group. Maurice, at Isidore's behest, did attend a good number of Illingworth, Morris board meetings, for the most part in place of Isidore himself. Despite the heady world of film production and the obvious leanings of Maurice and Mark toward this high-profile life, Isidore ensured that the more mundane world of producing cloth was not neglected and pursued a much more austere lifestyle.

The Ostrer family nicknamed Isidore 'Mephistopheles' or 'Mephi', and he was beyond doubt the most outstandingly intelligent of all six Ostrer children. The film actor James Mason – who was to marry his oldest daughter Pamela – said that the brothers had but one opinion and one brain between them (1996). Isidore was always in the lead in deciding where to invest money, in negotiating investments from others and in selecting loyal and capable managers in the widely different companies of which he had control. However, for the main part he led 'from behind', leaving his trusted brothers to the detail and rarely attending business meetings. While owning a country house in Berkshire, having a suite at Claridge's in later years and renting an apartment in Cannes, he was not noted for leading a lavish lifestyle, preferring to write poetry and books on economics above other pursuits.

During the period 1921 to 1939, despite the prevailing difficult context for the textile industry, Illingworth, Morris & Co. Ltd had continued to turn over adequate production levels and remain financially viable. Isidore's shareholdings in the group had been variable over these years, and shares were sold when money was required elsewhere but purchased when the market was favourable. His holdings in Illingworth, Morris were to vary between 23 per cent and 50 per cent over this period of time. On a personal note, Isidore's first marriage ended in divorce in 1931, and only four of the six children born by Helen Ostrer are registered in his surname. This suggests some impropriety by Helen and a reason for his taking action to divorce her, an action that was unusual for the times. He married for the second time on 27 April 1933 and his second wife Marjorie Roach had a chronic health problem, suffering from tuberculosis. Because of her condition Isidore moved firstly to Arizona and then to South Africa in the early 1940s, where the climate of both were thought helpful to her condition. In South Africa his Jewish background and left-wing, radical views made it difficult for him to condone a white supremacist society and, despite adopting a child (Isabella) there, he returned to live in England, leaving Marjorie behind. From then on, they met only occasionally in the south of France.

His progressive views extended from economics to health and lifestyle. He published *A New International Currency* (1921) and *The Conquest of Gold* (1923);

these were updated in 1964 with the publication *Modern Money and Unemployment*. The latter has a fulsome foreword written by Lord Beaverbrook, and is perhaps the clearest exposition of his theories. He held the view that unemployment was the greatest evil to society and that governments should strive to prevent this at all costs. The book also argues against retaining money in banks and for always using any money to produce goods and jobs. This philosophical perspective certainly affected his personal approach to his textile business and he became noted as 'always a buyer, never a seller' – a policy that built up some financial troubles for the years that lay ahead.

On his return to England in the 1940s, Isidore developed a pattern of wintering in his rented apartment in Cannes and spending the summer months in England – mostly at his Berkshire home. By this time, Maurice was also divorced and shared both the Cannes apartment and the Berkshire home with Isidore. Maurice was the brother who returned from Cannes in the winter to attend Illingworth, Morris board meetings and, from the 1950s, he truly was Isidore's shadow, performing necessary duties for him but appearing to have little personal control of the growing textile empire. In 1952, Isidore appointed Baron Wilmot of Selmeston as deputy chair of Illingworth Morris, and his brother Mark as a director – perhaps indicating that during his time away from England he had lost sight of some of the local, West Yorkshire talent within his large textile group of companies.

John Wilmot was born in London in 1893 and was educated at Hither Green School in Lewisham. From 1908, while working as an office boy, he had extended his education at Chelsea Polytechnic and later Kings College, London, where he was awarded the Gilbart prize in banking. He joined the Westminster bank, Piccadilly, while studying and rose rapidly to join the staff in the general manager's office, eventually becoming secretary to the Anglo-Russian bank. In 1909 he joined the Labour party and, after serving during the First World War in the Royal Naval Airforce, he helped to form the Lewisham Labour party. He was a candidate for this seat in the 1924 general election but narrowly failed to win the seat. After two more attempts to get elected he was successful in winning a seat at Fulham in a 1933 by-election. He was a passionate fighter against fascism in the 1930s and must have endeared himself to Isidore with his socialist beliefs and his hatred of Jewish persecution. In 1939 he was appointed to the National Executive Committee of the Labour Party, appointed a privy councillor in 1945 and awarded a baronetcy in 1950.

Isidore had astute understanding of character and left-wing views, and so, for a short time, the Illingworth, Morris board had been content to retain Sir Frank Sanderson as Chairman of Salts (Saltaire) after their acquisition of the company under Isidore's light-handed control. However, Sir Frank, a conservative, was replaced as chairman by Baron Wilmot of Selmeston in 1961. R. W. Guild and Robert Park Guild were initially retained at Salts, probably to sustain continuity, and were appointed as joint managing directors of Illingworth, Morris in 1961

(R. W. Guild was eighty-four years old at this point). Both father and son, who had presided over considerable acquisitions for Salts itself, must have had some difficulty in adjusting to the rather more impersonal demands from their new company board to adopt modern methods of management. They were perhaps doubtful that Illingworth, Morris would pay sufficient attention to Salts or care about its workforce in the intimate way that they had done. Illingworth, Morris directors began to require changes to purchasing and costing processes and introduced a more modern budgetary control costing process that began to identify some issues within the Salts company in relation to stock held and pricing cloth.

One of the authors of this book began researching in 2012 the personal histories of people who had worked at Salts from the 1930s onwards; one man, Frank Senior, who had started at Salts in the basement in 1933, packing cloth, provided some insight to the changes to practices for assessing the costs of cloth to be sold. Frank, who had gone to night school for qualifications, had been asked to take on the role as head of the costing department by R. W. Guild on his return from naval service in 1946, after the Second World War. He recalled that the method of costing in use from his starting work at Salts had been an Ernest H. Gates system, which had been adapted later to take on a John 'Wrights' and a 'Salts' (Saltaire) system. For some cloths a specific weaving company would be used – for example, men's suiting was woven at Salts Mill and ladies' cloth was woven by John Wrights or Ernest H. Gates at Crossroads, Keighley.

His department, having decided which companies' weaving costs they were going to use, then had to set the standard for other overheads (dyeing, finishing, warehousing and so on). Every year the accounts department would decide the overheads required, which were usually set at four times the weavers' wage, as these wages tended to be the highest. From this, Frank would give a price for different cloths to the Salts sales staff, who would often argue the price because Salts had some semi-automatic looms in New Mill that were less expensive to run. However, R. W. Guild always refuted this argument knowing the differentials relating to the cloth and weavers, which had been chosen in terms of quality. When the cloth was going out to Burtons (one of Salts' largest customers) it was essential to get the price right, and Frank remembered one occasion very well when R. W. asked his son Robert Park, 'How many thousands of yards do [Burtons] want?' Park's reply was, 'if we price anything above 20 shillings a yard, we won't get their order', and at this R. W. did compromise. The company benefited greatly from their relationship with Burtons for many years.

Illingworth, Morris, however, soon introduced 'budgetary control costing' and did away with Salts' 'plus four formulae'. For the new system, there had to be more staff, and all the separate processes were given their specific budgetary controls – for example, spinning cost controls and manufacturing cost controls – and each cost-control unit had to provide a monthly profit statement from their department. If the cloth was in a 'grey state', i.e., not finished, a notional finishing cost was added on. R. W. himself had to keep a record of all wool bought, what was in stock, what current prices were and what was an average raw material price. In addition, stock taking had occurred once a year in the time of the syndicate, but Illingworth, Morris wanted a stock take each month. It must have been difficult for a man who was as experienced as R. W. Guild, particularly at his great age, to implement these changes, albeit in a senior management capacity. R. W. Guild retired within three years of the takeover by Illingworth, Morris, and Park Guild also retired as managing director for Salts in 1962.

In other Illingworth, Morris companies, however, young talent was being identified and nurtured by the Ostrers. Donald Hanson's family had been discussing his further education on leaving school in 1939, aged fourteen, when the sudden death of the cashier at Globe Mills (Huddersfield) created an opportunity for Donald to apply for the job and continue his education at 'night school'. Donald's father, Harry, had been a manager at Globe for some time and had frequently taken his son to the mill at weekends and in the school holidays, explaining the machinery and the processes involved in manufacturing cloth, so Donald was no stranger to the complexities of the industry. After a break that involved service in the Royal Navy between 1943 and 1946, Donald returned to his job at Globe and continued his studies, completing a National Certificate in Commerce by 1948. He became chief cashier shortly afterwards and continued his studies to become an Associate of the Chartered Institute of Secretaries and Administrators – a highly valued commercial qualification. Donald recalled that Globe was a very successful company and not only produced first-class yarn but also good profit. It was a very

Donald Hanson (left) during armed service in the Second World War. (© Saltaire Archive)

welcome place in which to work in his view. He was appointed a director at Globe in 1955 and became managing director in 1958, aged thirty-three years.

There came a point in 1960 when it was necessary for him to leave Globe. Daniel Illingworth's, another company in the Illingworth, Morris group, was losing both money and their previously good reputation for high-quality white yarn. The board of Illingworth, Morris were aware of Donald's reputation for producing good yarn – gained in part through his father's knowledge and experience. In addition to his knowledge of the spinning process, he had also acquired a working knowledge of the preparation of raw wool. He had been fortunate to be employed in a company that produced its own tops and had been able to gain a lot of experience and knowledge about the early processes of washing and scouring wool. Daniel Illingworth's was about twice the size of Globe Mills but in 1960 was, according to Donald, in need of some 'tender loving care'. The chairman of the Illingworth, Morris group, Lord Wilmot, and the then board members asked Donald if he had sufficient time to carry out a survey of Whetley Mills (Daniel Illingworth's). Soon he was spending time at both Whetley Mills and at Globe; he discovered many flaws in the machinery at Whetley Mills and recommended a complete overhaul of certain sections of the work.

It took Donald just two weeks to carry out an investigation and he submitted his recommendations to the board as to what needed to be done. These included a recommendation to change the managing director then in post. This was welcomed, and he was asked to become the managing director of Daniel Illingworth's. He spent three years working there before all the changes had been completed and Daniel Illingworth's was back in profit. Having seemingly solved the difficulties there, the Illingworth, Morris board approached him again in 1964 and asked him to become managing director of James Tankard. James Tankard, the coloured spinners of Upper Croft Mills, Bradford, were carrying too much stock and this was, of course, detrimental to the balance sheet. Tankard were managed very well in their production methods but, to make the company profitable, it was essential to reduce their interest charges. The first priority was to make a very detailed examination of their stock of coloured tops. Donald discovered that too much stock in too many varieties of colour were being held and ensured that the stocks of coloured tops were reduced. It took three years to achieve changes but the balance sheet and profitability were restored by 1967.

By this time, Salts (Saltaire) Ltd had been notified that their director of combing and spinning, Douglas Rennie, had decided to retire. Donald had already been appointed a director of Salts and was asked to take over the combing and spinning department at Salts in 1967. Relations between the spinning and manufacturing departments were not good and there appeared to be a lack of cooperation between the process managers for both. He was determined to build good relationships with the director of the manufacturing department, Russell Viney. His efforts were successful and the relationship was on firm grounds by 1970, with

Salts profits increasing. Donald had become the man to fix internal problems in a number of the Illingworth, Morris group of companies. So much so that, in 1970, Isidore Ostrer, through his younger brother Maurice, wrote to the Illingworth, Morris board, instructing them to appoint Donald as a director of the Illingworth, Morris group. The board could not refuse the majority shareholder and appeared to welcome him with open arms, but Donald knew there was some disquiet about being ordered to do so.

It was in these years that Donald came to know Isidore and Maurice Ostrer very well. He met them on occasions in Claridge's, where Isidore had his own table and had persuaded the managers of the hotel to have lower lighting there at all times. Donald also spent time at Isidore's home at Titlarks Hill, Sunningdale, Berkshire, visiting on many occasions with his wife Margaret – always leaving there with a couple of bottles of fine wine. He recalls that Isidore, in a kindly way, was in full command of the giant textile company that Illingworth, Morris had become. He had great respect for both Isidore and Maurice, but found in Isidore a great charm and a deep appreciation of and concern for all the people who worked for him. The personal letters between Donald and Isidore Ostrer strongly indicate that, while

Donald Hanson, director of Illingworth, Morris.

Isidore may have been a financier first, he had social responsibility at the heart of his personality.

During the years that Donald Hanson was becoming known by Isidore and Maurice to be capable of overhauling elements of their vast business that were not performing well, the Illingworth, Morris group had carried on acquiring additional companies. For example, in 1954 they acquired a controlling interest in John Smith, Field Head Mills, spinners and combers. The group had become a very large one by the time of their acquisition of Salts (Saltaire) Ltd and their takeover of Salts in 1957 brought two giants of the textile industry together, forming a colossus. This, coupled with the Ostrers' ability to spot and promote talented men, had led to their not only being the single largest company in the industry from the 1960s onwards but to be in a position to record trading profits in 1964 of £2.08 million.

The Ostrers and their company board were not content to stop at that, however, and they acquired Bradford's largest woolcombing and top-making group, Woolcombers (Holdings) Ltd in December 1971.

The takeover of Woolcombers did not come as a surprise to the industry – both groups had exchanged directors in May of that year, with Donald Hanson of Illingworth, Morris being appointed to the board of Woolcombers and Alan Thompson of Woolcombers becoming vice chairman of Illingworth, Morris. Mark Keighley (2007) comments that it had been said that Illingworth, Morris had always had a grand plan to build the world's largest wool-textile company and it had taken many years for them to achieve it. This is a view that is perhaps not born out by their early approaches to their almost accidental textile acquisitions. Their achievements had been possible, in part, by buying companies and assets with money borrowed from the banks, and this had become an increasingly risky policy. By the mid-1970s, Illingworth, Morris's overdraft amounted to £26 million. Isidore's economic philosophy that regarded unemployment as the 'greatest evil in the world' had led to his rarely disposing of assets, his doing everything in his power to prevent closures and his being acutely aware of his responsibilities as an employer. While vastly different in style to Sir Titus Salt or Sir James Roberts, he nevertheless displayed – probably unconsciously – some similar traits in this respect and Salts Mill continued to benefit from a man who had more than business success as a personal motivation for his actions.

The risks taken by Isidore Ostrer were to be played out in full in the late 1970s. As principal shareholders of Illingworth, Morris & Co. Ltd, Isidore Ostrer died in September 1975, at his home in Sunningdale, and his brother Maurice died three months later in Cannes. Though somewhat different in character, they had nevertheless worked in a symbiotic relationship for many years and ended their days within months of each other. Isidore was eighty-six years old and Maurice seventy-nine years old. Their deaths left an intriguing stock market situation as between them they held about 53 per cent of the Illingworth, Morris equity capital. As Jack Wilks, editor of the *Wool Record,* commented,

> The deaths of the Ostrer brothers have pinpointed what must be a very rare situation in big company affairs – the destinies of not just Illingworth, Morris with its wide ranging interests in the wool-textile field, but of other important companies such as British Mohair Spinners and Hield Brothers, in which IM have built up a sizeable shareholding – are uncertain.

The person appointed as chief executrix of the estates of Isidore and Maurice Ostrer in 1976 was Pamela Mason. Pamela was Isidore's eldest daughter from his first marriage and had commenced a career as an actress when the Ostrers owned Gaumont-British. She married a cameraman, Roy Kellino, in her early teens but this marriage was short lived; she later married the British actor James Mason in 1940, with whom she had two children – a son, Morgan Mason, and a daughter, Portland – but this second marriage also ended in divorce. Her acting career was mostly in relatively minor films and she had moved to California, becoming a chat-show hostess and author there in the 1950s. She had property interests in Los Angeles and her first visit to Yorkshire created a ripple of excitement in the

Pamela Mason.

local press with headlines such as, 'Film Star keeps an eye on her shares'. She was, in fact, in control of 46 per cent of the Illingworth, Morris group shares, owning around 19 per cent of them herself but responsible for the remainder on behalf of other family legatees. She became a director of Illingworth, Morris in January 1976 and at the outset her relationship with other board members was amicable. She felt strongly that the group's products should be promoted more energetically overseas – particularly in the United States of America – views that the board felt to be pertinent and were happy to work on. At this point the company was performing well and she was too preoccupied with her life in Hollywood to take an active part in running the company; consequently, she appointed her son, Morgan, as a director of the company to represent her interests.

Morgan Mason made a brief tour of the Illingworth, Morris empire in the mid-1970s but was not able to become involved in its management, as he was heavily involved in the United States campaign to elect Ronald Reagan as President and eventually became a member of Reagan's staff in the White House. His mother, Pamela, was in no hurry to dispose of her stock, as the Illingworth, Morris Group were having a succession of good results and the business turnover rose to a record £118.93 million in 1977. The subsequent world-trade recession was to see the market for wool textiles collapse, causing the company to suffer substantial trading losses for the next two consecutive years. Under the chairmanship of Ivan Hill and subsequently of Donald Hanson, the group's directors worked effectively to reduce its overdraft by several millions; stock levels were reduced from £39 million to £25.7 million and the number of operating units were reduced from forty-two to twenty-six after full consultation with the trade unions, in cooperation with the staff and with the backing of the company's bankers.

These necessary and sensible moves to preserve the company's financial health did not meet with Pamela Mason's approval, however, and she began to intervene more and more in the board affairs. In 1979, she appointed Thomas Yeardye as international marketing director. Thomas Yeardye had been a boyfriend of the actress Diana Dors, had worked for Carmen Rollers and was vice president of Vidal Sassoon – not a background that would have provided insights into textile production. The board welcomed him but found a number of his ideas, such as his suggestion to build a carpet plant and produce Crombie products in the Nevada desert, to be unrealistic.

Pamela Mason became particularly unhappy with a deal proposed by the board with the group's bankers and, prior to the 1979 annual meeting of the board, she instructed the board's chairman, Ivan Hill, to remove all the British directors of the board. This was only withdrawn on the board's acceptance of Thomas Yeardye as a director and the removal of the company's auditors, Price Waterhouse, which satisfied her for a short time. Donald Hanson was appointed chairman of the group in August 1980 and as joint chief executive with Peter Hardy. Following Donald's appointment, Pamela proposed Casper Weinberger as a non-executive director for the company and this was agreed. When Casper Weinberger arrived in Saltaire. the 'Yorkshire' directors took an immediate liking to him and his appointment was unanimously approved. Unfortunately for the board, after only three months in this position, he was appointed US Secretary of State for Defence by the new President, Ronald Reagan, and could not continue as a board member. From this point on, Pamela Mason embarked on a process of attrition aimed at Illingworth, Morris board members, causing a stream of phone calls, faxes, letters and reports to pass between Bradford and Beverley Hills, coupled with constant threats to dismiss one or more of the directors. Peter Hardy, joint chief executive for the Illingworth, Morris group with Donald Hanson, arrived at Narita Airport, Tokyo, in February 1981 to be met by Thomas Yeardye, only to be told that he had been sacked.

Illingworth, Morris had been thrust into the media spotlight and had secured a very favourable deal with the group's bankers in 1980 that allowed for the company's debts to be restructured, with an agreed eighteen months' overdraft of £22 million. However, Pamela's opposition to 'giving the banks security', and her refusal to accept the deal, led the board of directors to agree it without her

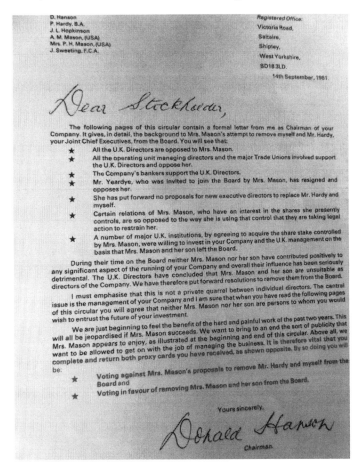

Donald Hanson's letter to shareholders, 1981.

approval. On 3 September 1981, Hill Samuel & Co., acting on behalf of the board, made an offer for the entire share capital controlled by Pamela (who at this point was holding off legal claims being made by other descendants of the Ostrers for their legacies) at a good market value, but she ignored the offer and it was withdrawn. As a result, an extraordinary general meeting of shareholders was called for 26 October 1981 to consider a proposal to remove Pamela and her son Morgan from the Illingworth, Morris board. The immensely detailed proposal, which runs to many pages, was to travel the length and breadth of the United Kingdom and to other countries where shareholders lived, so as to garner support for its contents. Pamela put forward a counter proposal to remove Donald Hanson and Peter Hardy as chairman and joint chief executives for the company. An original copy of the boards' proposal is held in the Saltaire Archive.

In the event, on 26 October 1981 the Masons were removed by 4,913,944 votes cast by 1,723 shareholders against 4,642,221 votes cast by twenty shareholders – Pamela Mason and nineteen others who controlled a further 2 per cent of the votes. The defeat of Pamela Mason set a record in stock-exchange history in that 95 per cent of all available votes had been cast. Illingworth, Morris & Co. Ltd immediately appointed three new directors – Geoffrey Kitchen, Jack Nunnerley and Sir Russell Sanderson – and Donald Hanson was able to announce that 'we have a chance to make the company what it should be – a viable and vital organisation'.

The company started to review its operations, selling a Scottish subsidiary company to Courtaulds and working to comply with the deal agreed in 1980 with the company's bankers. Pamela Mason, angry at her defeat, soon announced that she had agreed to sell her stake in the company to Abele, a Manx-based company controlled by Allan Lewis, for a lower price than that offered by Hill Samuel on behalf of the Illingworth, Morris board. She also commenced legal proceedings – serving writs and claims for damages on four of the Yorkshire directors – but these were never contested in court, being based on her views rather than any illegality. The headquarters at Salts Mill for Illingworth, Morris had never experienced events such as the ones that unfolded in the late 1970s and early 1980s, and the workers at Salts and residents of Saltaire could be forgiven for thinking that such a drama was more worthy of a Hollywood movie than the dry affairs that would be normal for senior management of a very important group of companies within the textile industrial community. As it was, the end of textile production at Salts Mill did not come about immediately after Pamela and Morgan Mason's removal. For four more years, in decreasing scale and range, textile production continued at Salts Mill and Illingworth, Morris eventually returned to being a profitable company.

Alan Lewis, who had agreed to buy Pamela Mason's 19 per cent personal stake in Illingworth, Morris, announced in 1982 that he was also about to acquire a further 27 per cent stake in the company that had been controlled by Pamela. When the deal was completed, he would have an interest in the company equivalent to that of Isidore Ostrer (46 per cent). His acquisition of this shareholding was referred to the Mergers and Monopolies commission, who ultimately judged that this would not be against the public interest and might prove valuable in restoring Illingworth, Morris. Some key groups representing the textile trade opposed the acquisition, but the opposition was short-lived and Alan Lewis became chairman of the Illingworth Morris group in 1983 – causing the *Wool Record* to comment that 'the sequel to a story of family conflict and industrial intrigue put the saga on a par with *Dallas*'.

Donald Hanson had remained for two more years as chairman of the company and negotiated an important deal with the British Wool Marketing Board during 1982. The outcome was an agreement for the Woolcombers subsidiary of Illingworth, Morris to sell 60 per cent of their shares in the merchanting division of Woolcombers to the British Wool Marketing Board; this gave Illingworth, Morris a £6 million improvement in their finances. By March 1982, group borrowings had reduced by a further £6 million and the balance sheet was greatly improved. This was largely due to the efforts of Donald Hanson and his fellow directors, and Donald retired from the company in 1983. Under the direction of Alan Lewis, by the end of March 1985, group borrowings were reduced to loans of £4.75 million by selling redundant and underperforming assets and, by 1987, Illingworth, Morris had an annual turnover of £96.2 million and were once again in the top league of the industry.

The future for Salts Mill itself was to be quite different. The textile industry was reducing in capacity and importance in Britain, and large mills such as Salts Mill were necessarily being restructured to reduce capacity greatly. In the early-to-mid 1980s, in order to look at innovative ways to utilise the space in the building, help and support was offered to small businesses to set up within the space available, action that resulted in a visit by the Prince of Wales's consort, Princess Diana. However, a future for Salts Mill in textile production became ever more problematic.

The then managing director of Salts, Mr Edward Stanners, commented in 1983, 'our building is vast, it was created for 4,000 workers, the heating of the mill is very expensive and while productivity is twice the volume achieved in 1979, the labour force has reduced by 40 per cent'. Illingworth, Morris announced in September 1985 that Salts Mill would cease production at the end of that year; spinning operations had been moved to Daniel Illingworth, Bradford, and the cloth manufacturing division and some of its assets were sold to Stroud Riley Drummond – including the trading name 'Salts (Saltaire) Ltd'. Alan Lewis, chairman and chief executive of the Illingworth, Morris group, had announced that year that, while the group's overall position had improved dramatically, there still remained a large amount of property that was superfluous to the group's manufacturing or trading needs.

Following a decision to close the Salts weaving operation, Illingworth, Morris then commenced a marketing operation designed to find a future for the giant mill complex.

Numerous options were explored, including the possibility of the beautiful old mill becoming a museum or the relocation of Shipley College to the building, and were the source of speculation for well over a year in the local press. John Collins, the head of the estates division for Illingworth, Morris, found a better solution early in 1987 and was largely responsible for the sale of the immense mill, intact and in good condition, to a young entrepreneur, Jonathan Silver. This marked the end of Sir Titus Salt's establishment of a grand, superbly designed mill for the purpose of worsted manufacture, which had lasted for 132 years, but it provided a fresh, innovative direction for the mill that has, to date, resulted in fully preserving the heritage that the building represents.

Postscript: Salts Mill 1987 to the Present

Salts Mill today has been very successfully regenerated by the late Jonathan Silver and his family. The vast mill has had its interior cleverly redesigned in a way that leaves the visitor with clear impressions of its former purpose. The mill offers a wide range of visually pleasing art, literature, specially themed exhibitions and retail outlets. Its success lies in the way in which each of the four floors of the mill, which are open to the public, have had the stone floors recovered, have superbly exposed brick and piping work (some painted in vivid colours) and creative art and photographic images adorning the walls. Each of these floors also house many other incidental items such as an amazing collection of Bermantofts pottery in vibrant shades. Much is displayed within those vast spaces: the books, the artists' materials, the designer homeware, the diner, an opera café, the locally crafted jewellery, the outdoor equipment outlet and antique dealers. These are spaces that are airy and light; feel welcoming and peaceful; contain large vases of fresh, scented flowers; and that display items in an innovative way that encourages exploration, holds the visitors' attention and stimulates their senses.

    The first collection of borrowed paintings from David Hockney that initiated the 1853 Gallery, on the first floor of the mill, has been added to many times and each of the many images and styles used by Hockney are displayed in a most attractive way. Today the mill exhibits the largest single collection of Hockney's work, frequently offering showings of his new work, and also displays the work of other artists. It has a range of very successful businesses as tenants, often involving the latest technology and scientific advancements in products. There are spaces dedicated for music, drama and literature events and spaces that offer insights into the mill's heritage – carefully planned among the whole. It is nothing short of a miracle of creativity and wonder, and has ensured the preservation of a beautifully designed mill for future generations. The mill today receives tens of thousands of visitors and is a major attraction for local residents, their families and guests and tourists from near and far.

This short postscript does not attempt to do justice to the Silver family who are the current owners of Salts Mill, and recommended reading for more detail about Jonathan Silver and his family is a book first published in 1997 by Jim Greenhalf, entitled *Salt and Silver, A Story of Hope*. This is now in its third edition and is published by Salts Mill Estates. Please do read this book in order to aid your understanding of the young entrepreneur who purchased the mill in 1987 and the many advantages that his vision and work, along with that of the Silver family, have brought to the village of Saltaire and their immense contribution to the conferring of World Heritage Status for Saltaire by UNESCO.

Bibliography

Baines, *Directory of the County of York* (1822).
Barlo and Shaw, *Balgarnie's Salt with commentary and additions* (2003).
Cabinet Papers, The National Archives, *The Economic Slump in the 1920s*.
Coates, C., *Genealogy of the Hanson Family* (2016).
Smith, M., *The Memories of Frank Senior 1918–1983* (2012).
Coates, C., *The Movements of Herbert Salt* (2016).
Coates, C., *Notes on George Salt* (2016).
Coates, C., *Notes on Sir James Hill & Timeline for Arthur James Hill, 1876–1935* (2016).
Coates, C., *Timeline for Baron Wilmot 1893–1964* (2016).
Coates, C., *Timeline for Charles Stead and Family* (2016).
Coates, C., *Timeline for Ernest Henry Gates, 1874–1925* (2016).
Coates, C., *Timeline for Isaac Smith* (2016).
Coates, C., *Timeline for Sir James Roberts* (2016).
Coates, C., *Timeline for John Maddocks* (2016).
Coates, C., *Timeline for John Rhodes* (2016).
Coates, C., *Timeline for Sir Frank Sanderson 1880–1965* (2016).
Coates, C., *Timeline for Sir Henry Whitehead* (2016).
Coates, C., *Timeline for Robert Whyte Guild 1876–1966* (2016).
Coates, C., *Timeline for Titus Salt Junior* (2016).
Frost, R., *A Brief History of Hawsons, The Sheffield Accountancy Firm* (2005).
Fox, F., *Wool, The Empire Industry* (1923).
Greenhalf, J., *Salt and Silver, A Story of Hope* (1997).
The Hanson Papers, The Original Annual meeting records for Amalgamated Textiles for 1919 and 1920.
James, D., *Oxford National Biography* (2004).

Keighley, M., *Wool City: A History of the Bradford Textile Industry in the 20th Century* (2007).

King, D., *Notes & Reproduction of the 'Statement and Scheme for Reconstruction', November 1892, Sir Titus Salt (Bart) & Sons and Co. Ltd* (2013).

King, D., *Notes on the Will and Codicils of the Late Sir Titus Salt, Bart* (2013).

King, D., *The Second Lord of Saltaire: The Family History of Sir James Roberts Bart. JP, LLD* (2012).

Lawson, T., *Edward Salt and Ferniehurst* (2016).

Lee-Van den Daele, R. and Beale, R., *Milner Field: The Lost Country House of Titus Salt Junior* (2013).

Ostrer, N., *The Ostrers and Gaumont British* (2010).

Reynolds, J., *The Great Paternalist, Titus Salt and the Growth of Nineteenth Century Bradford* (1983).

Thompson, D., *The Chartists* (1987).

Shaw, D. B., *Notes Taken from Documents held by Rhea County Archives for Dayton, Tennessee* (2012).

Smith, M., *Donald Hanson, An Extraordinary Life* (2013).

Smith, M., *Notes on the company listed as Sir Titus Salt (Bart) Sons and Co. Ltd which continued to be registered until 1929 when it went into liquidation* (2016). The West Yorkshire Archives.

Smith, M., *Notes from the Directors Minutes, Sir Titus Salt (Bart), Sons and Co. 1896–1918* (2016). The West Yorkshire Archives.

Smith, M., *Notes from the Roberts Family Collection* (2016).

Smith, M., *Notes of Interviews with John Collins* (2012).

Watson, I., *Lecture Notes, The Letters of Sir James Roberts, 1914–1917* (2016).

Wells, D. A., *Recent Economic Changes and Their Effect on Production and Distribution of Wealth and Well-Being of Society* (1890).

Wright, E. M., *The Life of Joseph Wright* (1932).

A considerable amount of new research undertaken by Colin Coates for this book, from January to June 2016, has been drawn from the British Library, British and international newspaper archives; Ancestry.co.uk and relevant census returns for the period. National newspaper and international newspaper archive sources include: *The Bradford Observer, Leeds Times, Leeds Mercury, Bradford Daily Telegraph, Yorkshire Post, Yorkshire Gazette, Huddersfield Chronicle, Halifax Courier, Yorkshire Evening Post, Evening Freeman, Dublin, Lancaster Gazette, Western Times, Illustrated Times, London Daily News* and *New York Times*, and other local newspapers where the event or individual has been reported. This has resulted in a number of specific timelines that are now available in the Saltaire Archive.

Some of the processes involved in manufacture of worsted cloth

Sorting: Using experience to sort the wool fleeces manually into defined qualities.

Blending: Combining wools of similar qualities but different types to obtain some desired effect and produce a bulk lot for combing.

Scouring: Washing the wool to remove dirt, natural grease and any other impurities.

Carding: Disentangling the wool fibres so these can be made into a twist, less rope-like form, called a sliver.

Backwashing: A light washing of the slivers to remove dirt picked up in carding.

Combing: By machine after the mid-1880s, to straighten the wool fibres and separate the short wool from the long.

Drawing: Pulling and refining the combed tops until they are reduced from thick slivers of wool to roving from which the yarn is finally spun.

Spinning: Spinning the wool, drawing it out to its final thickness and twisting it for added strength.

Twisting: Taking two or more single spun yarns to produce a yarn of greater strength, for use as warp threads in the weaving process.

Reeling: The spun yarns are formed into hanks for dyeing.

Winding: Placing spun yarns into packages.

Warping: A large number of ends are wound, side by side, in predetermined order, density and width, onto a beam for the loom.

Warp Preparation: Individual warp threads are drawn through shafts according to a pattern set by designers.

Weaving: Where a loom is used to produce a piece of cloth by interfacing the warp threads (running the length of the fabric) with the weft threads (passing from side to side). The warp threads by use of a warping beam and the weft by use of a shuttle.

<u>Perching</u>: The woven fabric is inspected for faults and these are marked.

<u>Burling & Mending</u>: The meticulous process of taking out minor faults and rectifying these by hand.

<u>Dyeing</u>: A wet process in which raw materials, yarns and fabrics have colour applied.

<u>Finishing</u>: The fabric is treated by various processes to produce the required effect, feel and handling quality.

Salts, being a 'vertical mill', carried out most or all of the above processes for the majority of its history in textile manufacture.